George A. Anastassiou and Oktay Duman

Towards Intelligent Modeling: Statistical Approximation Theory

T0181325

Intelligent Systems Reference Library, Volume 14

Editors-in-Chief

Prof. Janusz Kacprzyk
Systems Research Institute
Polish Academy of Sciences
ul. Newelska 6
01-447 Warsaw
Poland
E-mail: kacprzyk@ibspan.waw.pl

Prof. Lakhmi C. Jain
University of South Australia
Adelaide
Mawson Lakes Campus
South Australia 5095
Australia
E-mail: Lakhmi.jain@unisa.edu.au

Further volumes of this series can be found on our
homepage: springer.com

George A. Anastassiou and Oktay Duman

Towards Intelligent Modeling: Statistical Approximation Theory

 Springer

George A. Anastassiou
The University of Memphis
Department of Mathematical Sciences
Memphis, TN 38152
USA
E-mail: ganastss@gmail.com

Oktay Duman
TOBB Economics and Technology University
Department of Mathematics
Sögütözü
TR-06530, Ankara
Turkey

ISBN 978-3-642-26817-5   ISBN 978-3-642-19826-7 (eBook)

DOI 10.1007/978-3-642-19826-7

Intelligent Systems Reference Library   ISSN 1868-4394

Typeset & *Cover Design:* Scientific Publishing Services Pvt. Ltd., Chennai, India.

Printed on acid-free paper

9 8 7 6 5 4 3 2 1

springer.com

Dedicated to *Angela, Bahar, Doğa, Esra, Koula, Peggy*

and also

dedicated to **World Peace!**

George Anastassiou in Iguazu Falls, Brazil, March 2008.

Oktay Duman, Ankara, Turkey, January 2011.

Preface

In the classical convergence analysis, almost all terms of a sequence have to belong to arbitrarily small neighborhood of the limit. The main idea of statistical convergence, which was first introduced by Fast in 1952 (see [70]), is to relax this condition and to demand validity of the convergence condition only for a majority of elements. This method of convergence has been investigated in many fundamental areas of mathematics, such as, measure theory, approximation theory, fuzzy logic theory, summability theory, and so on. These studies demonstrate that the concept of statistical convergence provides an important contribution to improvement of the classical analysis. In this book, we mainly consider this concept in approximation a function by linear operators, especially, when the classical limit fails. The results in the book cover not only the classical and statistical approximation theorems but also many significant applications to the fuzzy logic theory with the help of fuzzy-valued operators.

The study of the Korovkin-type approximation theory is an area of active research, which deals with the problem of approximating a function by means of a sequence of positive linear operators. Recently, this theory has been improved in two directions. The first one is *the statistical Korovkin theory*, which was first considered by Gadjiev and Orhan in 2002 (see [80]); and the second one is *the fuzzy Korovkin theory* introduced by the first author of this book in 2005 (see [6]). The main idea of this book is to combine these directions and is to present many significant applications. In this book, we also give various statistical approximation theorems for some specific (real or complex-valued) linear operators which do not need to be positive.

This is the first monograph in Statistical Approximation Theory and Fuzziness, which contains mostly the authors' joint research works on these topics of the last five years. Chapters are self-contained and several advanced courses can be taught out of this book.

We display lots of applications but always within the framework of Statistical Approximation. A complete list of references is presented at the end. In Chapter 1, we collect some necessary materials about Statistical Convergence and also Fuzzy Real Analysis, which provides a background for interested readers. In Chapters 2-5 we present many statistical approximation results with applications for some specific linear operators which do not need to be positive, such as, bivariate Picard and Gauss-Weierstrass operators. In Chapter 6 we introduce a Baskakov-type generalization of the Statistical Korovkin Theory. In Chapter 7, we mainly prove that it is possible to approximate in statistical sense to derivatives of functions defined on weighted spaces. In Chapter 8, we obtain some statistical approximation results in trigonometric case. In Chapter 9, we present various results relaxing the positivity conditions of statistical approximating operators. In Chapters 10, 11 we obtain statistical Korovkin-type approximation theorems for univariate and multivariate stochastic processes. In Chapters 12, 13 we present fractional Korovkin results based on statistical convergence in algebraic and trigonometric cases. In Chapters 14-16 we introduce many fuzzy statistical approximation theorems for fuzzy positive linear operators. In Chapters 17, 18 we present statistical convergence of bivariate complex Picard and Gauss-Weierstrass integral operators.

This monograph is suitable for graduate students and researchers in pure and applied mathematics and engineering, it is great for seminars and advanced graduate courses, also to be in all science and engineering libraries.

We would like to thank our families for their patience and support.

November 2010

George A. Anastassiou
Department of Mathematical Sciences
The University of Memphis, TN, USA

Oktay Duman
TOBB Economics and Technology University
Department of Mathematics, Ankara, Turkey

Contents

1

Introduction

In this chapter, we mainly collect all necessary materials used in this book and discuss their fundamental properties. We also give some brief descriptions of the chapters.

1.1 Background and Preliminaries

The notion of statistical convergence, while introduced over nearly fifty years ago (see [70, 73]), has only recently become an area of active research. Different mathematicians studied properties of statistical convergence and applied this concept in various areas, such as, measure theory [72, 75, 98, 99]; trigonometric series [108]; locally convex spaces [96]; summability theory [46, 47, 71, 92]; densities of subsets of the natural numbers [101]; the Stone-Čhech compactification of the natural numbers [50]; Banach spaces [48]; number sequences [45, 49, 52, 74, 76, 106]; the fuzzy set theory [44, 90, 102]; and so on. This is because it is quite effective, especially, when the classical limit of a sequence fails. As usual, according to the ordinary convergence, almost all elements of a sequence have to belong to arbitrarily small neighborhood of the limit; but the main idea of statistical convergence is to relax this condition and to demand validity of the convergence condition only for a majority of elements. Statistical convergence which is a regular non-matrix summability method is also effective in summing non-convergent sequences. Recent studies demonstrate that the notion of statistical convergence provides an important contribution to

G.A. Anastassiou and O. Duman: Towards Intelligent Modeling, ISRL 14, pp. 1–8.
springerlink.com   © Springer-Verlag Berlin Heidelberg 2011

improvement of the classical analysis. Furthermore, in recent years, this convergence method has been used in the approximation theory settings, which is known as *statistical approximation theory* in the literature.

The classical Korovkin theory (see, e.g., [1, 53, 93]), which is one of the most familiar area in the approximation theory, is mainly based on two conditions: the *positivity* of linear operators and the *validity* of their (ordinary) limits. However, we know that these conditions can be weakened. For the first situation, complex-valued or fuzzy-valued operators are used in general (see [4, 5, 10, 36, 78, 82, 83, 87, 95]); but there are also many real-valued approximating operators that are not positive, such as, Picard, Poisson-Cauchy and Gauss-Weierstrass singular integral operators (see [7–9, 32, 34, 35, 37–39, 81]). For the second situation, especially, the concept of *statistical convergence* from the summability theory plays a crucial role so that it is possible to approximate (in statistical sense) a function by means of a sequence of linear operators although the limit of the sequence fails (see [16–31, 54–65, 67–69, 80, 89, 104, 105]).

The main purpose of this book is to present the recent developments on the statistical approximation theory.

First of all, we give some basic definitions and notations on the concept of statistical convergence.

Definition 1.1 (see [101, 108]). *Let K be a subset of $\mathbb{N}$, the set of all natural numbers. Then, the (asymptotic) density of K, denoted by $\delta(K)$, is given by*

$$\delta(K) := \lim_j \frac{1}{j} \# \{n \le j : n \in K\}$$

whenever the limit exists, where $\#B$ denotes the cardinality of the set B.

According to this definition, there are also some subsets of $\mathbb{N}$ having no density. For example, let A be the set of all even positive integers, B_1 the set of all even positive integers with an even number of digits to base ten, and B_2 the set of all odd positive integers with an odd number of digits. Define $B = B_1 \cup B_2$. Then, observe that $\delta(A \cup B)$ and $\delta(A \cap B)$ do not exist (see, e.g., [101, p. 248]).

On the other hand, it is easy to check the following facts:

- $\delta(\mathbb{N}) = 1$,

- $\delta(\{2n : n \in \mathbb{N}\}) = \delta(\{2n - 1 : n \in \mathbb{N}\}) = 1/2$,

- $\delta(\{n^2 : n \in \mathbb{N}\}) = 0$,

- $\delta(\{n : n \text{ is prime}\}) = 0$,

- if K is a finite subset of $\mathbb{N}$, then $\delta(K) = 0$,

- if $\delta(K)$ exists, then $\delta(\mathbb{N} - K) = 1 - \delta(K)$,

- if $K_1 \subset K_2$ and $\delta(K_1)$, $\delta(K_2)$ exist, then $0 \leq \delta(K_1) \leq \delta(K_2) \leq 1$,

- if $\delta(K_1) = \delta(K_2) = 1$, then $\delta(K_1 \cup K_2) = \delta(K_1 \cap K_2) = 1$,

- if $\delta(K_1) = \delta(K_2) = 0$, then $\delta(K_1 \cup K_2) = \delta(K_1 \cap K_2) = 0$.

Using this density, Fast [70] introduced the concept of statistical convergence as follows.

Definition 1.2 (see [70]). *A number sequence* (x_n) *is statistically convergent to* L *if, for every* $\varepsilon > 0$,

$$\delta(\{n : |x_n - L| \geq \varepsilon\}) = 0,$$

or, equivalently,

$$\lim_{j} \frac{1}{j} \# \{n \leq j : |x_n - L| \geq \varepsilon\} = 0$$

for every $\varepsilon > 0$. *In this case, we write* $st - \lim_n x_n = L$.

We immediately obtain that every convergent sequence is statistically convergent to the same value, but the converse is not always true. Not all properties of convergent sequences hold true for statistical convergence. For instance, although it is well-known that a subsequence of a convergent sequence is convergent, this is not always true for statistical convergence. Another example is that every convergent sequence must be bounded, however it does not need to be bounded of an statistically convergent sequence.

The following definition weakens the boundedness of a sequence (see, e.g., [52, 76]).

Definition 1.3 (see [52, 76]). *A sequence* (x_n) *is called statistically bounded if there exists a subset* K *of* $\mathbb{N}$ *with density one such that, for every* $n \in K$, $|x_n| \leq M$ *holds for some positive constant* M.

Then, one can see that every bounded sequence is statistically bounded but not conversely, and also that a statistical convergent sequence must be statistically bounded. Connor [45] proved the following useful characterization for statistical convergence.

Theorem 1.4 (see [45]). $st - \lim_n x_n = L$ *if and only if there exists an index set* K *with* $\delta(K) = 1$ *such that* $\lim_{n \in K} x_n = L$, *i.e., for every* $\varepsilon > 0$, *there is a number* $n_0 \in K$ *such that* $|x_n - L| < \varepsilon$ *holds for all* $n \geq n_0$ *with* $n \in K$.

Now define the sequence (x_n) by

$$x_n := \begin{cases} \sqrt{n}, \text{ if } n = m^2, \ m \in \mathbb{N} \\ 0, \quad \text{otherwise.} \end{cases}$$

Then observe that (x_n) is unbounded above and so divergent; but it statistically converges to 0. This simple example explains the power of statistical convergence.

Now let $A := [a_{jn}]$, $j, n = 1, 2, ...$, be an infinite summability matrix. Then, the following definition is well-known in the summability theory.

Definition 1.5 (see [42, 86]). *For a given sequence (x_n), the A- transform of x, denoted by $((Ax)_j)$, is given by*

$$(Ax)_j = \sum_{n=1}^{\infty} a_{jn} x_n$$

provided the series converges for each $j \in \mathbb{N}$. We say that A is regular if

$$\lim_j (Ax)_j = L \quad \text{whenever} \quad \lim_n x_n = L.$$

The next characterization regarding the regularity of a matrix A is known in the literature as the Silverman-Toeplitz conditions.

Theorem 1.6 (see [42, 86]). *An infinite summability matrix $A = [a_{jn}]$ is regular if and only if it satisfies all of the following properties*

- $\sup_j \sum_{n=1}^{\infty} |a_{jn}| < \infty$,
- $\lim_j a_{jn} = 0$ *for each $n \in \mathbb{N}$,*
- $\lim_j \sum_{n=1}^{\infty} a_{jn} = 1$.

For example, the well-known regular summability matrix is the Cesáro matrix $C_1 = [c_{jn}]$ given by

$$c_{jn} := \begin{cases} \dfrac{1}{j}, & \text{if } 1 \leq n \leq j \\ 0, & \text{otherwise.} \end{cases}$$

since the absolute row sums are bounded, every column sequence converges to 0, and the row sums converge to 1.

Using the regular matrices, Freedman and Sember [71] extent the statistical convergence to the concept of A-statistical convergence by the following way.

Let $A = [a_{jn}]$ be a non-negative regular summability matrix.

Definition 1.7 (see [71]). *The A-density of a subset K of $\mathbb{N}$ is defined by*

$$\delta_A (K) = \lim_j \sum_{n \in K} a_{jn}$$

provided that the limit exists.

If we take $A = C_1$, then we have $\delta_{C_1}(K) = \delta(K)$.

Definition 1.8 (see [71]). *A sequence (x_n) is called A-statistically convergent to L if, for every $\varepsilon > 0$,*

$$\delta_A(\{n : |x_n - L| \geq \varepsilon\}) = 0,$$

or, equivalently,

$$\lim_j \sum_{n \,:\, |x_n - L| \geq \varepsilon} a_{jn} = 0.$$

This limit is denoted by $st_A - \lim_n x_n = L$.

Of course, if we take $A = C_1$, then C_1-statistical convergence coincides with statistical convergence. Also, observe that if $A = I$, the identity matrix, then we get the ordinary convergence. It is clear that every convergent sequences is A-statistically convergent, however the converse is not always true. Actually, if $A = [a_{jn}]$ is any non-negative regular summability matrix satisfying the condition

$$\lim_j \max_n \{a_{jn}\} = 0,$$

then A-statistical convergence method is stronger than convergence (see [92]). We should note that Theorem 1.4 is also valid for A-statistical convergence (see [99]).

We now focus on the fuzzy theory.

Definition 1.9. *A fuzzy number is a function $\mu : \mathbb{R} \to [0,1]$, which is normal, convex, upper semi-continuous and the closure of the set $supp(\mu)$ is compact, where $supp(\mu) := \{x \in \mathbb{R} : \mu(x) > 0\}$. The set of all fuzzy numbers are denoted by $\mathbb{R}_{\mathcal{F}}$.*

Let

$$[\mu]^0 := \overline{\{x \in \mathbb{R} : \mu(x) > 0\}} \text{ and } [\mu]^r := \{x \in \mathbb{R} : \mu(x) \geq r\}, \ (0 < r \leq 1).$$

Then, it is well-known [84] that, for each $r \in [0,1]$, the set $[\mu]^r$ is a closed and bounded interval of $\mathbb{R}$. For any $u, v \in \mathbb{R}_{\mathcal{F}}$ and $\lambda \in \mathbb{R}$, it is possible to define uniquely the sum $u \oplus v$ and the product $\lambda \odot u$ as follows:

$$[u \oplus v]^r = [u]^r + [v]^r \text{ and } [\lambda \odot u]^r = \lambda[u]^r, \ (0 \leq r \leq 1).$$

Now denote the interval $[u]^r$ by $[u_-^{(r)}, u_+^{(r)}]$, where $u_-^{(r)} \leq u_+^{(r)}$ and $u_-^{(r)}, u_+^{(r)} \in \mathbb{R}$ for $r \in [0,1]$. Then, for $u, v \in \mathbb{R}_{\mathcal{F}}$, define

$$u \preceq v \Leftrightarrow u_-^{(r)} \leq v_-^{(r)} \text{ and } u_+^{(r)} \leq v_+^{(r)} \text{ for all } 0 \leq r \leq 1.$$

Define also the following metric $D : \mathbb{R}_{\mathcal{F}} \times \mathbb{R}_{\mathcal{F}} \to \mathbb{R}_{+}$ by

$$D(u, v) = \sup_{r \in [0,1]} \max \left\{ \left| u_-^{(r)} - v_-^{(r)} \right|, \left| u_+^{(r)} - v_+^{(r)} \right| \right\}.$$

In this case, $(\mathbb{R}_{\mathcal{F}}, D)$ is a complete metric space (see [107]) with the properties

$$D(u \oplus w, v \oplus w) = D(u, v) \quad \text{for all } u, v, w \in \mathbb{R}_{\mathcal{F}},$$
$$D(\lambda \odot u, \lambda \odot v) = |\lambda| \, D(u, v) \quad \text{for all } u, v \in \mathbb{R}_{\mathcal{F}} \text{ and } \lambda \in \mathbb{R},$$
$$D(u \oplus v, w \oplus z) \leq D(u, w) + D(v, z) \quad \text{for all } u, v, w, z \in \mathbb{R}_{\mathcal{F}}.$$

Let $f, g : [a, b] \to \mathbb{R}_{\mathcal{F}}$ be fuzzy number valued functions. Then, the distance between f and g is given by

$$D^*(f, g) = \sup_{x \in [a,b]} \sup_{r \in [0,1]} \max \left\{ \left| f_-^{(r)} - g_-^{(r)} \right|, \left| f_+^{(r)} - g_+^{(r)} \right| \right\}.$$

Nuray and Savaş [102] introduced the fuzzy analog of statistical convergence by using the metric D as the following way.

Definition 1.10 (see [102]). *Let (μ_n) be a fuzzy number valued sequence. Then, (μ_n) is called statistically convergent to a fuzzy number μ if, for every $\varepsilon > 0$,*

$$\lim_j \frac{\# \{ n \leq j : D(\mu_n, \mu) \geq \varepsilon \}}{j} = 0$$

holds. This limit is denoted by $st - \lim_n D(\mu_n, \mu) = 0$.

Assume now that $A = [a_{jn}]$ is a non-negative regular summability matrix. Then, the above definition can easily be generalized, the so-called A-statistical convergence, as follows:

Definition 1.11. *A fuzzy valued sequence (μ_n) is A-statistically convergent to $\mu \in \mathbb{R}_{\mathcal{F}}$, which is denoted by $st_A - \lim_n D(\mu_n, \mu) = 0$, if for every $\varepsilon > 0$,*

$$\lim_j \sum_{n : D(\mu_n, \mu) \geq \varepsilon} a_{jn} = 0$$

holds.

It is not hard to see that in the case of $A = C_1$, Definition 1.11 reduces to Definition 1.10. Furthermore, if A is replaced by the identity matrix, then we have the fuzzy convergence introduced by Matloka (see [97]).

1.2 Chapters Description

In this section, we describe our monograph's chapters.

In Chapter 2, we construct a sequence of bivariate smooth Picard singular integral operators which do not have to be positive in general. After giving some useful estimates, we mainly prove that it is possible to approximate a function by these operators in statistical sense even though they do not obey the positivity condition of the statistical Korovkin theory.

In Chapter 3, we study the statistical approximation properties of a sequence of bivariate smooth Gauss-Weierstrass singular integral operators which are not positive in general.

In Chapter 4, we obtain some statistical approximation results for the bivariate smooth Picard singular integral operators defined on L_p-spaces. Also, giving a non-trivial example we show that the statistical L_p-approximation is more applicable than the ordinary one.

In Chapter 5, we study statistical L_p-approximation properties of the bivariate Gauss-Weierstrass singular integral operators. Furthermore, we introduce a non-trivial example showing that the statistical L_p-approximation is more powerful than the ordinary case.

In Chapter 6, with the help of the notion of A-statistical convergence, where A is a non-negative regular summability matrix, we get some statistical variants of Baskakov's results on the Korovkin-type approximation theorems.

In Chapter 7, we prove some Korovkin-type approximation theorems providing the statistical weighted convergence to derivatives of functions by means of a class of linear operators acting on weighted spaces. We also discuss the contribution of these results to the approximation theory.

In Chapter 8, using A-statistical convergence and also considering some matrix summability methods, we introduce an approximation theorem, which is a non-trivial generalization of Baskakov's result regarding the approximation to periodic functions by a general class of linear operators.

In Chapter 9, we relax the positivity condition of linear operators in the Korovkin-type approximation theory via the concept of statistical convergence. Especially, we prove some Korovkin-type approximation theorems providing the statistical convergence to derivatives of functions by means of a class of linear operators.

In Chapter 10, we present strong Korovkin-type approximation theorems for stochastic processes via the concept of A-statistical convergence.

In Chapter 11, we obtain some Korovkin-type approximation theorems for multivariate stochastic processes with the help of the concept of A-statistical convergence. A non-trivial example showing the importance of this method of approximation is also introduced.

In Chapter 12, we give some statistical Korovkin-type approximation theorems including fractional derivatives of functions. We also demonstrate that these results are more applicable than the classical ones.

In Chapter 13, we develop the classical trigonometric Korovkin theory by using the concept of statistical convergence from the summability theory and also by considering the fractional derivatives of trigonometric functions.

In Chapter 14, we present a Korovkin-type approximation theorem for fuzzy positive linear operators by using the notion of A-statistical convergence. This type of approximation enables us to obtain more powerful results than in the classical aspects of approximation theory settings. An application of this result is also presented. Furthermore, we study the rates of this statistical fuzzy convergence of the operators via the fuzzy modulus of continuity.

In Chapter 15, we consider non-negative regular summability matrix transformations in the approximation by fuzzy positive linear operators, where the test functions are trigonometric. So, we mainly obtain a trigonometric fuzzy Korovkin theorem by means of A-statistical convergence. We also compute the rates of A-statistical convergence of a sequence of fuzzy positive linear operators in the trigonometric environment.

In Chapter 16, we obtain a statistical fuzzy Korovkin-type approximation result with high rate of convergence. Main tools used in this work are statistical convergence and higher order continuously differentiable functions in the fuzzy sense. An application is also given, which demonstrates that the statistical fuzzy approximation is stronger than the classical one.

In Chapter 17, we investigate some statistical approximation properties of the bivariate complex Picard integral operators. Furthermore, we show that the statistical approach is more applicable than the well-known aspects.

In Chapter 18, we present the complex Gauss-Weierstrass integral operators defined on a space of analytic functions in two variables on the Cartesian product of two unit disks. Then, we investigate some geometric properties and statistical approximation process of these operators.

2

Statistical Approximation by Bivariate Picard Singular Integral Operators

At first we construct a sequence of bivariate smooth Picard singular integral operators which do not have to be positive in general. After giving some useful estimates, we mainly prove that it is possible to approximate a function by these operators in statistical sense even though they do not obey the positivity condition of the statistical Korovkin theory. This chapter relies on [25].

2.1 Definition of the Operators

Throughout this section, for $r \in \mathbb{N}$ and $m \in \mathbb{N}_0 := \mathbb{N} \cup \{0\}$, we use

$$
\alpha_{j,r}^{[m]} := \begin{cases} (-1)^{r-j} \binom{r}{j} j^{-m} & \text{if } j = 1, 2, ..., r, \\ 1 - \sum\limits_{j=1}^{r} (-1)^{r-j} \binom{r}{j} j^{-m} & \text{if } j = 0. \end{cases} \tag{2.1}
$$

and

$$
\delta_{k,r}^{[m]} := \sum_{j=1}^{r} \alpha_{j,r}^{[m]} j^k, \quad k = 1, 2, ..., m \in \mathbb{N}. \tag{2.2}
$$

Then observe that

$$
\sum_{j=0}^{r} \alpha_{j,r}^{[m]} = 1 \tag{2.3}
$$

G.A. Anastassiou and O. Duman: Towards Intelligent Modeling, ISRL 14, pp. 9–23.
springerlink.com

and

$$-\sum_{j=1}^{r}(-1)^{r-j}\binom{r}{j} = (-1)^r\binom{r}{0}. \qquad (2.4)$$

We now define the bivariate smooth Picard singular integral operators as follows:

$$P_{r,n}^{[m]}(f;x,y) = \frac{1}{2\pi\xi_n^2}\sum_{j=0}^{r}\alpha_{j,r}^{[m]}\left(\int_{-\infty}^{\infty}\int_{-\infty}^{\infty} f\left(x+sj,y+tj\right)e^{-(\sqrt{s^2+t^2})/\xi_n}dsdt\right),$$
$$(2.5)$$

where $(x,y) \in \mathbb{R}^2$, $n,r \in \mathbb{N}$, $m \in \mathbb{N}_0$, $f : \mathbb{R}^2 \to \mathbb{R}$ is a Lebesgue measurable function, and also (ξ_n) is a bounded sequence of positive real numbers.

We make

Remark 2.1. *The operators $P_{r,n}^{[m]}$ are not in general positive. For example, consider the function $\varphi(u,v) = u^2 + v^2$ and also take $r = 2$, $m = 3$, $x = y = 0$. Observe that $\varphi \geq 0$, however*

$$P_{2,n}^{[3]}(\varphi;0,0) = \frac{1}{2\pi\xi_n^2}\left(\sum_{j=1}^{2}j^2\alpha_{j,2}^{[3]}\right)\int_{-\infty}^{\infty}\int_{-\infty}^{\infty}\left(s^2+t^2\right)e^{-(\sqrt{s^2+t^2})/\xi_n}dsdt$$

$$= \frac{2}{\pi\xi_n^2}\left(\alpha_{1,2}^{[3]}+4\alpha_{2,2}^{[3]}\right)\int_{0}^{\infty}\int_{0}^{\infty}\left(s^2+t^2\right)e^{-(\sqrt{s^2+t^2})/\xi_n}dsdt$$

$$= \frac{2}{\pi\xi_n^2}\left(-2+\frac{1}{2}\right)\int_{0}^{\pi/2}\int_{0}^{\infty}e^{-\rho/\xi_n}\rho^3 d\rho d\theta$$

$$= -9\xi_n^2 < 0.$$

We need

Lemma 2.2. *The operators $P_{r,n}^{[m]}$ given by (2.5) preserve the constant functions in two variables.*

Proof. Let $f(x, y) = C$, where C is any real constant. By (2.1) and (2.3), we obtain, for every $r, n \in \mathbb{N}$ and $m \in \mathbb{N}_0$, that

$$
P_{r,n}^{[m]}(C; x, y) = \frac{C}{2\pi \xi_n^2} \sum_{j=0}^{r} \alpha_{j,r}^{[m]} \left(\int_{-\infty}^{\infty} \int_{-\infty}^{\infty} e^{-(\sqrt{s^2+t^2})/\xi_n} \, ds \, dt \right)
$$

$$
= \frac{C}{2\pi \xi_n^2} \int_{-\infty}^{\infty} \int_{-\infty}^{\infty} e^{-(\sqrt{s^2+t^2})/\xi_n} \, ds \, dt
$$

$$
= \frac{2C}{\pi \xi_n^2} \int_{0}^{\infty} \int_{0}^{\infty} e^{-(\sqrt{s^2+t^2})/\xi_n} \, ds \, dt
$$

$$
= \frac{2C}{\pi \xi_n^2} \int_{0}^{\pi/2} \int_{0}^{\infty} e^{-\rho/\xi_n} \rho \, d\rho \, d\theta
$$

$$
= \frac{C}{\xi_n^2} \int_{0}^{\infty} e^{-\rho/\xi_n} \rho \, d\rho
$$

$$
= C,
$$

which finishes the proof. ∎

We also need

Lemma 2.3. *Let $k \in \mathbb{N}_0$. Then, it holds, for each $\ell = 0, 1, ..., k$ and for every $n \in \mathbb{N}$, that*

$$
\int_{-\infty}^{\infty} \int_{-\infty}^{\infty} s^{k-\ell} t^\ell e^{-(\sqrt{s^2+t^2})/\xi_n} \, ds \, dt
$$

$$
= \begin{cases} 2B\left(\frac{k-\ell+1}{2}, \frac{\ell+1}{2}\right) \xi_n^{k+2}(k+1)! & \text{if } k \text{ and } \ell \text{ are even} \\ 0 & \text{otherwise,} \end{cases}
$$

where $B(a, b)$ denotes the Beta function.

Proof. It is clear that if k or ℓ are odd, then the integrand is a odd function with respect to s and t; and hence the above integral is zero. Also, if both k and ℓ are even, then the integrand is an even function with respect to s

and t. So, we derive that

$$
\int\limits_{-\infty}^{\infty} \int\limits_{-\infty}^{\infty} s^{k-\ell} t^{\ell} e^{-(\sqrt{s^2+t^2})/\xi_n} ds dt = 4 \int\limits_{0}^{\infty} \int\limits_{0}^{\infty} s^{k-\ell} t^{\ell} e^{-(\sqrt{s^2+t^2})/\xi_n} ds dt
$$

$$
= 4 \int\limits_{0}^{\pi/2} (\cos\theta)^{k-\ell} (\sin\theta)^{\ell} d\theta \int\limits_{0}^{\infty} \rho^{k+1} e^{-\rho/\xi_n} d\rho
$$

$$
= 2B\left(\frac{k-\ell+1}{2}, \frac{\ell+1}{2}\right) \xi_n^{k+2} (k+1)!
$$

proving the result. ∎

2.2 Estimates for the Operators

Let $f \in C_B(\mathbb{R}^2)$, the space of all bounded and continuous functions on $\mathbb{R}^2$. Then, the rth (bivariate) modulus of smoothness of f is given by (see, e.g., [33])

$$
\omega_r(f; h) := \sup_{\sqrt{u^2+v^2}\leq h} \left\| \Delta_{u,v}^r(f) \right\| < \infty, \quad h > 0, \tag{2.6}
$$

where $\|\cdot\|$ is the sup-norm and

$$
\Delta_{u,v}^r (f(x,y)) = \sum_{j=0}^{r} (-1)^{r-j} \binom{r}{j} f(x + ju, y + jv). \tag{2.7}
$$

Let $m \in \mathbb{N}$. By $C^{(m)}(\mathbb{R}^2)$ we denote the space of functions having m times continuous partial derivatives with respect to the variables x and y. Assume now that a function $f \in C^{(m)}(\mathbb{R}^2)$ satisfies the condition

$$
\left\| \frac{\partial^m f(\cdot, \cdot)}{\partial^{m-\ell} x \partial^\ell y} \right\| := \sup_{(x,y)\in\mathbb{R}^2} \left| \frac{\partial^m f(x,y)}{\partial^{m-\ell} x \partial^\ell y} \right| < \infty \tag{2.8}
$$

for every $\ell = 0, 1, ..., m$. Then, we consider the function

$$
G_{x,y}^{[m]}(s,t) := \frac{1}{(m-1)!} \sum_{j=0}^{r} \binom{r}{j} \int\limits_0^1 (1-w)^{m-1}
$$
$$
\times \left\{ \sum_{\ell=0}^{m} \binom{m}{m-\ell} \left| \frac{\partial^m f(x + jsw, y + jtw)}{\partial^{m-\ell} x \partial^\ell y} \right| \right\} dw \tag{2.9}
$$

for $m \in \mathbb{N}$ and $(x,y), (s,t) \in \mathbb{R}^2$. Notice that the condition (2.8) implies that $G_{x,y}^{[m]}(s,t)$ is well-defined for each fixed $m \in \mathbb{N}$.

We initially estimate the case of $m \in \mathbb{N}$ in (2.5).

Theorem 2.4. *Let $m \in \mathbb{N}$ and $f \in C^{(m)}\left(\mathbb{R}^2\right)$ for which (2.8) holds. Then, for the operators $P_{r,n}^{[m]}$, we have*

$$
\left| P_{r,n}^{[m]}(f;x,y) - f(x,y) - \frac{1}{\pi} \sum_{i=1}^{[m/2]} (2i+1)\delta_{2i,r}^{[m]} \xi_n^{2i} \right.
$$

$$
\left. \times \left\{ \sum_{\ell=0}^{2i} \binom{2i}{2i-\ell} \frac{\partial^{2i} f(x,y)}{\partial^{2i-\ell}x \partial^\ell y} B\left(\frac{2i-\ell+1}{2}, \frac{\ell+1}{2}\right) \right\} \right| \qquad (2.10)
$$

$$
\leq \frac{1}{2\pi \xi_n^2} \int_{-\infty}^{\infty} \int_{-\infty}^{\infty} G_{x,y}^{[m]}(s,t)(|s|^m + |t|^m) e^{-(\sqrt{s^2+t^2})/\xi_n} \, ds\, dt.
$$

The sums in the left hand side of (2.10) collapse when $m = 1$.

Proof. Let $(x,y) \in \mathbb{R}^2$ be fixed. By Taylor's formula, we have that

$$
f(x+js, y+jt) = \sum_{k=0}^{m-1} \frac{j^k}{k!} \sum_{\ell=0}^{k} \binom{k}{k-\ell} s^{k-\ell} t^\ell \frac{\partial^k f(x,y)}{\partial^{k-\ell}x \partial^\ell y}
$$

$$
+ \frac{j^m}{(m-1)!} \int_0^1 (1-w)^{m-1} \left\{ \sum_{\ell=0}^{m} \binom{m}{m-\ell} s^{m-\ell} t^\ell \right.
$$

$$
\left. \times \frac{\partial^m f(x+jsw, y+jtw)}{\partial^{m-\ell}x \partial^\ell y} \right\} dw.
$$

The last implies

$$
f(x+js, y+jt) - f(x,y) = \sum_{k=1}^{m} \frac{j^k}{k!} \sum_{\ell=0}^{k} \binom{k}{k-\ell} s^{k-\ell} t^\ell \frac{\partial^k f(x,y)}{\partial^{k-\ell}x \partial^\ell y}
$$

$$
- \frac{j^m}{(m-1)!} \int_0^1 (1-w)^{m-1}
$$

$$
\times \left\{ \sum_{\ell=0}^{m} \binom{m}{m-\ell} s^{m-\ell} t^\ell \frac{\partial^m f(x,y)}{\partial^{m-\ell}x \partial^\ell y} \right\} dw
$$

$$
+ \frac{j^m}{(m-1)!} \int_0^1 (1-w)^{m-1} \left\{ \sum_{\ell=0}^{m} \binom{m}{m-\ell} s^{m-\ell} t^\ell \right.
$$

$$
\left. \times \frac{\partial^m f(x+jsw, y+jtw)}{\partial^{m-\ell}x \partial^\ell y} \right\} dw.
$$

Now multiplying both sides of the above equality by $\alpha_{j,r}^{[m]}$ and summing up from 0 to r we get

$$
\sum_{j=0}^{r} \alpha_{j,r}^{[m]} \left(f(x+js, y+jt) - f(x,y) \right) = \sum_{k=1}^{m} \frac{\delta_{k,r}^{[m]}}{k!} \sum_{\ell=0}^{k} \binom{k}{k-\ell} s^{k-\ell} t^{\ell} \frac{\partial^{k} f(x,y)}{\partial^{k-\ell} x \partial^{\ell} y}
$$
$$
+ \frac{1}{(m-1)!} \int_{0}^{1} (1-w)^{m-1} \varphi_{s,t}^{[m]}(w) dw,
$$

where

$$
\varphi_{x,y}^{[m]}(w; s, t) = \sum_{j=0}^{r} \alpha_{j,r}^{[m]} j^{m} \left\{ \sum_{\ell=0}^{m} \binom{m}{m-\ell} s^{m-\ell} t^{\ell} \frac{\partial^{m} f(x+jsw, y+jtw)}{\partial^{m-\ell} x \partial^{\ell} y} \right\}
$$
$$
- \delta_{m,r}^{[m]} \sum_{\ell=0}^{m} \binom{m}{m-\ell} s^{m-\ell} t^{\ell} \frac{\partial^{m} f(x,y)}{\partial^{m-\ell} x \partial^{\ell} y}.
$$

At first we estimate $\varphi_{x,y}^{[m]}(w; s, t)$. Using (2.1), (2.2) and (2.4), we have

$$
\varphi_{x,y}^{[m]}(w; s, t) = \sum_{j=1}^{r} (-1)^{r-j} \binom{r}{j} \left\{ \sum_{\ell=0}^{m} \binom{m}{m-\ell} s^{m-\ell} t^{\ell} \frac{\partial^{m} f(x+jsw, y+jtw)}{\partial^{m-\ell} x \partial^{\ell} y} \right\}
$$
$$
- \sum_{j=1}^{r} (-1)^{r-j} \binom{r}{j} \left\{ \sum_{\ell=0}^{m} \binom{m}{m-\ell} s^{m-\ell} t^{\ell} \frac{\partial^{m} f(x,y)}{\partial^{m-\ell} x \partial^{\ell} y} \right\}
$$
$$
= \sum_{j=1}^{r} (-1)^{r-j} \binom{r}{j} \left\{ \sum_{\ell=0}^{m} \binom{m}{m-\ell} s^{m-\ell} t^{\ell} \frac{\partial^{m} f(x+jsw, y+jtw)}{\partial^{m-\ell} x \partial^{\ell} y} \right\}
$$
$$
+ (-1)^{r} \binom{r}{0} \left\{ \sum_{\ell=0}^{m} \binom{m}{m-\ell} s^{m-\ell} t^{\ell} \frac{\partial^{m} f(x,y)}{\partial^{m-\ell} x \partial^{\ell} y} \right\}
$$
$$
= \sum_{j=0}^{r} (-1)^{r-j} \binom{r}{j} \left\{ \sum_{\ell=0}^{m} \binom{m}{m-\ell} s^{m-\ell} t^{\ell} \frac{\partial^{m} f(x+jsw, y+jtw)}{\partial^{m-\ell} x \partial^{\ell} y} \right\}.
$$

In this case, we observe that

$$
\left| \varphi_{x,y}^{[m]}(w; s, t) \right| \leq (|s|^{m} + |t|^{m}) \sum_{j=0}^{r} \binom{r}{j} \left\{ \sum_{\ell=0}^{m} \binom{m}{m-\ell} \left| \frac{\partial^{m} f(x+jsw, y+jtw)}{\partial^{m-\ell} x \partial^{\ell} y} \right| \right\}.
$$
$$
\tag{2.11}
$$

After integration and some simple calculations, and also using Lemma 2.2, we find, for every $n \in \mathbb{N}$, that

$$
\begin{aligned}
P_{r,n}^{[m]}(f; x, y) - f(x, y) &= \frac{1}{2\pi\xi_n^2} \int\limits_{-\infty}^{\infty} \int\limits_{-\infty}^{\infty} \left\{ \sum_{j=0}^{r} \alpha_{j,r}^{[m]} \left(f\left(x + sj, y + tj\right) - f(x, y) \right) \right\} \\
&\quad \times e^{-(\sqrt{s^2+t^2})/\xi_n} ds dt \\
&= \frac{1}{2\pi\xi_n^2} \sum_{k=1}^{m} \frac{\delta_{k,r}^{[m]}}{k!} \sum_{\ell=0}^{k} \binom{k}{k-\ell} \frac{\partial^k f(x, y)}{\partial^{k-\ell}x \partial^\ell y} \\
&\quad \times \left\{ \int\limits_{-\infty}^{\infty} \int\limits_{-\infty}^{\infty} s^{k-\ell} t^\ell e^{-(\sqrt{s^2+t^2})/\xi_n} ds dt \right\} \\
&\quad + R_n^{[m]}(x, y)
\end{aligned}
$$

with

$$
R_n^{[m]}(x, y) := \frac{1}{2\pi\xi_n^2 (m-1)!} \int\limits_{-\infty}^{\infty} \int\limits_{-\infty}^{\infty} \left(\int\limits_0^1 (1-w)^{m-1} \varphi_{x,y}^{[m]}(w; s, t) dw \right) e^{-(\sqrt{s^2+t^2})/\xi_n} ds dt.
$$

By (2.9) and (2.11), it is clear that

$$
\left| R_n^{[m]}(x, y) \right| \leq \frac{1}{2\pi\xi_n^2} \int\limits_{-\infty}^{\infty} \int\limits_{-\infty}^{\infty} G_{x,y}^{[m]}(s, t) \left(|s|^m + |t|^m \right) e^{-(\sqrt{s^2+t^2})/\xi_n} ds dt.
$$

Then, combining these results with Lemma 2.3, we immediately derive (2.10). The proof is completed. ∎

The next estimate answers the case of $m = 0$ in (2.5).

Theorem 2.5. *Let $f \in C_B\left(\mathbb{R}^2\right)$. Then, we have*

$$
\left| P_{r,n}^{[0]}(f; x, y) - f(x, y) \right| \leq \frac{2}{\pi\xi_n^2} \int\limits_0^{\infty} \int\limits_0^{\infty} \omega_r \left(f; \sqrt{s^2 + t^2} \right) e^{-(\sqrt{s^2+t^2})/\xi_n} ds dt.
$$

$$
(2.12)
$$

Proof. Taking $m = 0$ in (2.1) we notice that

$$
P_{r,n}^{[0]}(f;x,y) - f(x,y) = \frac{1}{2\pi\xi_n^2} \int\limits_{-\infty}^{\infty} \int\limits_{-\infty}^{\infty} \left\{ \sum_{j=1}^{r} \alpha_{j,r}^{[0]} \left(f\left(x+sj,y+tj\right) - f(x,y)\right) \right\}
$$
$$
\times e^{-(\sqrt{s^2+t^2})/\xi_n}\,ds dt
$$
$$
= \frac{1}{2\pi\xi_n^2} \int\limits_{-\infty}^{\infty} \int\limits_{-\infty}^{\infty} \left\{ \sum_{j=1}^{r} (-1)^{r-j} \binom{r}{j} \left(f\left(x+sj,y+tj\right) - f(x,y)\right) \right\}
$$
$$
\times e^{-(\sqrt{s^2+t^2})/\xi_n}\,ds dt
$$
$$
= \frac{1}{2\pi\xi_n^2} \int\limits_{-\infty}^{\infty} \int\limits_{-\infty}^{\infty} \left\{ \sum_{j=1}^{r} (-1)^{r-j} \binom{r}{j} f\left(x+sj,y+tj\right) \right.
$$
$$
\left. + \left(-\sum_{j=1}^{r} (-1)^{r-j} \binom{r}{j} \right) f(x,y) \right\} e^{-(\sqrt{s^2+t^2})/\xi_n}\,ds dt.
$$

Now employing (2.4) we have

$$
P_{r,n}^{[0]}(f;x,y) - f(x,y) = \frac{1}{2\pi\xi_n^2} \int\limits_{-\infty}^{\infty} \int\limits_{-\infty}^{\infty} \left\{ \sum_{j=1}^{r} (-1)^{r-j} \binom{r}{j} f\left(x+sj,y+tj\right) \right.
$$
$$
\left. + (-1)^{r} \binom{r}{0} f(x,y) \right\} e^{-(\sqrt{s^2+t^2})/\xi_n}\,ds dt
$$
$$
= \frac{1}{2\pi\xi_n^2} \int\limits_{-\infty}^{\infty} \int\limits_{-\infty}^{\infty} \left\{ \sum_{j=0}^{r} (-1)^{r-j} \binom{r}{j} f\left(x+sj,y+tj\right) \right\}
$$
$$
\times e^{-(\sqrt{s^2+t^2})/\xi_n}\,ds dt,
$$

and hence, by (2.7),

$$
P_{r,n}^{[0]}(f;x,y) - f(x,y) = \frac{1}{2\pi\xi_n^2} \int\limits_{-\infty}^{\infty} \int\limits_{-\infty}^{\infty} \Delta_{s,t}^{r} \left(f(x,y)\right) e^{-(\sqrt{s^2+t^2})/\xi_n}\,ds dt.
$$

Therefore, we derive from (2.6) that

$$
\begin{aligned}
\left| P_{r,n}^{[0]}(f; x, y) - f(x, y) \right| &\leq \frac{1}{2\pi \xi_n^2} \int_{-\infty}^{\infty} \int_{-\infty}^{\infty} \left| \Delta_{s,t}^r \left(f(x, y) \right) \right| e^{-(\sqrt{s^2+t^2})/\xi_n} \, ds \, dt \\
&\leq \frac{1}{2\pi \xi_n^2} \int_{-\infty}^{\infty} \int_{-\infty}^{\infty} \omega_r \left(f; \sqrt{s^2+t^2} \right) e^{-(\sqrt{s^2+t^2})/\xi_n} \, ds \, dt \\
&= \frac{2}{\pi \xi_n^2} \int_0^{\infty} \int_0^{\infty} \omega_r \left(f; \sqrt{s^2+t^2} \right) e^{-(\sqrt{s^2+t^2})/\xi_n} \, ds \, dt
\end{aligned}
$$

which establishes the proof.   ∎

2.3 Statistical Approximation of the Operators

We first obtain the following statistical approximation theorem for the operators (2.5) in case of $m \in \mathbb{N}$.

Theorem 2.6. *Let $A = [a_{jn}]$ be a non-negative regular summability matrix, and let (ξ_n) be a bounded sequence of positive real numbers for which*

$$
st_A - \lim_n \xi_n = 0 \tag{2.13}
$$

holds. Then, for each fixed $m \in \mathbb{N}$ and for all $f \in C^{(m)}\left(\mathbb{R}^2\right)$ satisfying (2.8), we have

$$
st_A - \lim_n \left\| P_{r,n}^{[m]}(f) - f \right\| = 0. \tag{2.14}
$$

Proof. Let $m \in \mathbb{N}$ be fixed. Then, we get from the hypothesis and (2.10) that

$$
\begin{aligned}
\left\| P_{r,n}^{[m]}(f) - f \right\| &\leq \sum_{i=1}^{[m/2]} (2i+1) K_i \delta_{2i,r}^{[m]} \xi_n^{2i} \\
&+ \frac{1}{2\pi \xi_n^2} \int_{-\infty}^{\infty} \int_{-\infty}^{\infty} \left\| G_{x,y}^{[m]}(s,t) \right\| (|s|^m + |t|^m) e^{-(\sqrt{s^2+t^2})/\xi_n} \, ds \, dt,
\end{aligned}
$$

where

$$
K_i := \frac{1}{\pi} \sum_{\ell=0}^{2i} \binom{2i}{2i-\ell} \left\| \frac{\partial^{2i} f(\cdot, \cdot)}{\partial^{2i-\ell} x \partial^\ell y} \right\| B \left(\frac{2i-\ell+1}{2}, \frac{\ell+1}{2} \right)
$$

for $i = 1, ..., \left[\frac{m}{2}\right]$. By (2.9) we obtain that

$$\left\|G_{x,y}^{[m]}(s,t)\right\| \leq \frac{2^r}{(m-1)!} \left(\sum_{\ell=0}^{m} \binom{m}{m-\ell} \left\|\frac{\partial^m f(\cdot,\cdot)}{\partial^{m-\ell}x \partial^\ell y}\right\|\right) \int_0^1 (1-w)^{m-1} dw$$

$$= \frac{2^r}{m!} \sum_{\ell=0}^{m} \binom{m}{m-\ell} \left\|\frac{\partial^m f(\cdot,\cdot)}{\partial^{m-\ell}x \partial^\ell y}\right\|,$$

thus we derive

$$\left\|P_{r,n}^{[m]}(f) - f\right\| \leq \sum_{i=1}^{[m/2]} (2i+1)K_i \delta_{2i,r}^{[m]} \xi_n^{2i}$$

$$+ \frac{2^{r+1}}{\pi m! \xi_n^2} \left(\sum_{\ell=0}^{m} \binom{m}{m-\ell} \left\|\frac{\partial^m f(\cdot,\cdot)}{\partial^{m-\ell}x \partial^\ell y}\right\|\right)$$

$$\times \int_0^\infty \int_0^\infty (s^m + t^m) e^{-(\sqrt{s^2+t^2})/\xi_n} ds dt.$$

Hence, we have

$$\left\|P_{r,n}^{[m]}(f) - f\right\| \leq \sum_{i=1}^{[m/2]} (2i+1)K_i \delta_{2i,r}^{[m]} \xi_n^{2i}$$

$$+ L_m \int_0^\infty \int_0^\infty (s^m + t^m) e^{-(\sqrt{s^2+t^2})/\xi_n} ds dt$$

$$= \sum_{i=1}^{[m/2]} (2i+1)K_i \delta_{2i,r}^{[m]} \xi_n^{2i}$$

$$+ L_m \int_0^{\pi/2} \int_0^\infty (\cos^m \theta + \sin^m \theta) \rho^{m+1} e^{-\rho/\xi_n} d\rho d\theta,$$

where

$$L_m := \frac{2^{r+1}}{\pi m! \xi_n^2} \left(\sum_{\ell=0}^{m} \binom{m}{m-\ell} \left\|\frac{\partial^m f(\cdot,\cdot)}{\partial^{m-\ell}x \partial^\ell y}\right\|\right).$$

After some simple calculations, we observe that

$$\left\|P_{r,n}^{[m]}(f) - f\right\| \leq \sum_{i=1}^{[m/2]} (2i+1)K_i \delta_{2i,r}^{[m]} \xi_n^{2i} + L_m \xi_n^{m+2}(m+1)! U_m,$$

where

$$U_m := \int_0^{\pi/2} (\cos^m \theta + \sin^m \theta) d\theta = B\left(\frac{m+1}{2}, \frac{1}{2}\right),$$

which yields

$$\left\| P_{r,n}^{[m]}(f) - f \right\| \le S_m \left\{ \xi_n^{m+2} + \sum_{i=1}^{[m/2]} \xi_n^{2i} \right\}, \tag{2.15}$$

with

$$S_m := (m+1)! U_m L_m + \max_{i=1,2,\dots,[m/2]} \left\{ (2i+1) K_i \delta_{2i,r}^{[m]} \right\}.$$

Next, for a given $\varepsilon > 0$, define the following sets:

$$D := \left\{ n \in \mathbb{N} : \left\| P_{r,n}^{[m]}(f) - f \right\| \ge \varepsilon \right\},$$

$$D_i := \left\{ n \in \mathbb{N} : \xi_n^{2i} \ge \frac{\varepsilon}{(1 + [m/2]) S_m} \right\}, \quad i = 1, \dots, \left[\frac{m}{2} \right],$$

$$D_{1+[m/2]} := \left\{ n \in \mathbb{N} : \xi_n^{m+2} \ge \frac{\varepsilon}{(1 + [m/2]) S_m} \right\}.$$

Then, the inequality (2.15) gives that

$$D \subseteq \bigcup_{i=1}^{1+[m/2]} D_i,$$

and thus, for every $j \in \mathbb{N}$,

$$\sum_{n \in D} a_{jn} \le \sum_{i=1}^{1+[m/2]} \sum_{n \in D_i} a_{jn}.$$

Now taking limit as $j \to \infty$ in the both sides of the above inequality and using the hypothesis (2.13), we obtain that

$$\lim_j \sum_{n \in D} a_{jn} = 0,$$

which implies (2.14). So, the proof is done. ∎

Finally, we investigate the statistical approximation properties of the operators (2.5) when $m = 0$. We need the following result.

Lemma 2.7. *Let $A = [a_{jn}]$ be a non-negative regular summability matrix, and let (ξ_n) be a bounded sequence of positive real numbers for which (2.13) holds. Then, for every $f \in C_B\left(\mathbb{R}^2\right)$, we have*

$$st_A - \lim_n \omega_r \left(f; \xi_n \right) = 0. \tag{2.16}$$

Proof. By the right-continuity of $\omega_r \left(f; \cdot \right)$ at zero, we may write that, for a given $\varepsilon > 0$, there exists a $\delta > 0$ such that $\omega_r \left(f; h \right) < \varepsilon$ whenever

$0 < h < \delta$. Hence, $\omega_r(f;h) \geq \varepsilon$ implies that $h \geq \delta$. Now replacing h by ξ_n, for every $\varepsilon > 0$, we see that

$$\{n : \omega_r(f;\xi_n) \geq \varepsilon\} \subseteq \{n : \xi_n \geq \delta\},$$

which guarantees that, for each $j \in \mathbb{N}$,

$$\sum_{n:\omega_r(f;\xi_n)\geq\varepsilon} a_{jn} \leq \sum_{n:\xi_n\geq\delta} a_{jn}.$$

Also, by (2.13), we obtain

$$\lim_j \sum_{n:\xi_n\geq\delta} a_{jn} = 0,$$

which implies

$$\lim_j \sum_{n:\omega_r(f;\xi_n)\geq\varepsilon} a_{jn} = 0.$$

So, the proof is finished.   ∎

Theorem 2.8. *Let* $A = [a_{jn}]$ *be a non-negative regular summability matrix, and let* (ξ_n) *be a bounded sequence of positive real numbers for which* *(2.13) holds. Then, for all* $f \in C_B(\mathbb{R}^2)$, *we have*

$$st_A - \lim_n \left\| P_{r,n}^{[0]}(f) - f \right\| = 0. \tag{2.17}$$

Proof. By (2.12), we can write

$$\left\| P_{r,n}^{[0]}(f) - f \right\| \leq \frac{2}{\pi\xi_n^2} \int_0^\infty \int_0^\infty \omega_r\left(f; \sqrt{s^2+t^2}\right) e^{-(\sqrt{s^2+t^2})/\xi_n} \, ds \, dt.$$

Now using the fact that $\omega_r\left(f;\lambda u\right) \le (1+\lambda)^r \omega_r\left(f;u\right),\ \lambda, u > 0$, we obtain

$$\left\|P_{r,n}^{[0]}(f) - f\right\| \le \frac{2}{\pi\xi_n^2} \int_0^\infty \int_0^\infty \omega_r\left(f;\xi_n\frac{\sqrt{s^2+t^2}}{\xi_n}\right) e^{-(\sqrt{s^2+t^2})/\xi_n}\,ds\,dt$$

$$\le \frac{2\omega_r\left(f;\xi_n\right)}{\pi\xi_n^2} \int_0^\infty \int_0^\infty \left(1 + \frac{\sqrt{s^2+t^2}}{\xi_n}\right)^r e^{-(\sqrt{s^2+t^2})/\xi_n}\,ds\,dt$$

$$= \frac{2\omega_r\left(f;\xi_n\right)}{\pi\xi_n^2} \int_0^{\pi/2} \int_0^\infty \left(1 + \frac{\rho}{\xi_n}\right)^r \rho e^{-\rho/\xi_n}\,d\rho\,d\theta$$

$$= \omega_r\left(f;\xi_n\right) \int_0^\infty \left(1+u\right)^r ue^{-u}\,du$$

$$\le \omega_r\left(f;\xi_n\right) \int_0^\infty \left(1+u\right)^{r+1} e^{-u}\,du$$

$$= \left(\sum_{k=0}^{r+1} \binom{r+1}{k}k!\right) \omega_r\left(f;\xi_n\right).$$

So that

$$\left\|P_{r,n}^{[0]}(f) - f\right\| \le K_r\omega_r\left(f;\xi_n\right), \tag{2.18}$$

where

$$K_r := \sum_{k=0}^{r+1} \binom{r+1}{k}k!.$$

Then, from (2.18), for a given $\varepsilon > 0$, we see that

$$\left\{n \in \mathbb{N} : \left\|P_{r,n}^{[0]}(f) - f\right\| \ge \varepsilon\right\} \subseteq \left\{n \in \mathbb{N} : \omega_r\left(f;\xi_n\right) \ge \frac{\varepsilon}{K_r}\right\},$$

which implies that

$$\sum_{n:\left\|P_{r,n}^{[0]}(f)-f\right\|\ge\varepsilon} a_{jn} \le \sum_{n:\omega_r(f;\xi_n)\ge\varepsilon/K_r} a_{jn} \tag{2.19}$$

holds for every $j \in \mathbb{N}$. Now, taking limit as $j \to \infty$ in the both sides of inequality (2.19) and also using (2.16), we get that

$$\lim_j \sum_{n:\left\|P_{r,n}^{[0]}(f)-f\right\|\ge\varepsilon} a_{jn} = 0,$$

which means (2.17). Hence, the proof is completed. ∎

2.4 Conclusions

In this section, we give some special cases of the results presented in the previous sections.

Taking $A = C_1$, the Cesáro matrix of order one, and also combining Theorems 2.6 and 2.8, we immediately derive the following result.

Corollary 2.9. *Let (ξ_n) be a bounded sequence of positive real numbers for which*

$$st - \lim_n \xi_n = 0$$

holds. Then, for each fixed $m \in \mathbb{N}_0$ and for all $f \in C^{(m)}\left(\mathbb{R}^2\right)$ satisfying (2.8), we have

$$st - \lim_n \left\| P_{r,n}^{[m]}(f) - f \right\| = 0.$$

Furthermore, choosing $A = I$, the identity matrix, in Theorems 2.6 and 2.8, we get the next approximation theorems with the usual convergence.

Corollary 2.10. *Let (ξ_n) be a bounded sequence of positive real numbers for which*

$$\lim_n \xi_n = 0$$

holds. Then, for each fixed $m \in \mathbb{N}_0$ and for all $f \in C^{(m)}\left(\mathbb{R}^2\right)$ satisfying (2.8), the sequence $\left(P_{r,n}^{[m]}(f)\right)$ is uniformly convergent to f on $\mathbb{R}^2$.

Next we define a special sequence (ξ_n) as follows:

$$\xi_n := \begin{cases} 1, & \text{if } n = k^2, \ k = 1, 2, \dots \\ \frac{1}{n}, & \text{otherwise.} \end{cases} \tag{2.20}$$

Then, observe that $st - \lim_n \xi_n = 0$. In this case, taking $A = C_1$, we get from Corollary 2.9 (or, Theorems 2.6 and 2.8) that

$$st - \lim_n \left\| P_{r,n}^{[m]}(f) - f \right\| = 0$$

holds for each $m \in \mathbb{N}_0$ and for all $f \in C^{(m)}\left(\mathbb{R}^2\right)$ satisfying (2.8). However, since the sequence (ξ_n) given by (2.20) is non-convergent, the classical approximation to a function f by the operators $P_{r,n}^{[m]}(f)$ is impossible.

Notice that Theorems 2.6, 2.8 and Corollary 2.9 are also valid when $\lim \xi_n = 0$ because every convergent sequence is A-statistically convergent, and so statistically convergent. But, as in the above example, the theorems obtained in this chapter still work although (ξ_n) is non-convergent. Therefore, this non-trivial example clearly demonstrates the power of the statistical approximation method in Theorems 2.6 and 2.8 with respect to Corollary 2.10.

At the end, we should remark that, usually, almost all statistical approximation results deal with positive linear operators. Of course, in this case, one has the following natural question:

- Can we use the concept of A-statistical convergence in the approximation by non-positive approximation operators?

The same question was also asked as an open problem by Duman et. al. in [62]. In this chapter we give affirmative answers to this problem by using the bivariate smooth Picard singular integral operators given by (2.5). However, some similar arguments may be valid for other non-positive operators.

3

Uniform Approximation in Statistical Sense by Bivariate Gauss-Weierstrass Singular Integral Operators

In this chapter, we study the statistical approximation properties of a sequence of bivariate smooth Gauss-Weierstrass singular integral operators which are not positive in general. We also show that the statistical approximation results are stronger than the classical uniform approximations. This chapter relies on [28].

3.1 Definition of the Operators

In this section we introduce a sequence of bivariate smooth Gauss-Weierstrass singular integral operators. We first give some notation used in the chapter. Let

$$
\alpha_{j,r}^{[m]} := \begin{cases} (-1)^{r-j} \binom{r}{j} j^{-m} & \text{if } j = 1, 2, ..., r, \\ 1 - \sum_{j=1}^{r} (-1)^{r-j} \binom{r}{j} j^{-m} & \text{if } j = 0. \end{cases} \tag{3.1}
$$

and

$$
\delta_{k,r}^{[m]} := \sum_{j=1}^{r} \alpha_{j,r}^{[m]} j^k, \quad k = 1, 2, ..., m \in \mathbb{N}. \tag{3.2}
$$

Then it is clear that $\sum_{j=0}^{r} \alpha_{j,r}^{[m]} = 1$ and $-\sum_{j=1}^{r} (-1)^{r-j} \binom{r}{j} = (-1)^r \binom{r}{0}$ hold. We also observe the set

$$
\mathbb{D} := \left\{ (s, t) \in \mathbb{R}^2 : s^2 + t^2 \leq \pi^2 \right\}.
$$

G.A. Anastassiou and O. Duman: Towards Intelligent Modeling, ISRL 14, pp. 25–38.
springerlink.com

Assume now that (ξ_n) is a sequence of positive real numbers. Setting

$$\lambda_n := \frac{1}{\pi\left(1 - e^{-\pi^2/\xi_n^2}\right)}, \tag{3.3}$$

we define the bivariate smooth Gauss-Weierstrass singular integral operators as follows:

$$W_{r,n}^{[m]}(f;x,y) = \frac{\lambda_n}{\xi_n^2} \sum_{j=0}^{r} \alpha_{j,r}^{[m]} \left(\iint_{\mathbb{D}} f\left(x+sj, y+tj\right) e^{-(s^2+t^2)/\xi_n^2} dsdt \right), \tag{3.4}$$

where $(x,y) \in \mathbb{D}$, $n,r \in \mathbb{N}$, $m \in \mathbb{N}_0 := \mathbb{N} \cup \{0\}$, and also $f : \mathbb{D} \to \mathbb{R}$ is a Lebesgue measurable function. In this case, we see that the operators $W_{r,n}^{[m]}$ are not positive in general. For example, if we take $\varphi(u,v) = u^2 + v^2$ and also take $r = 2$, $m = 3$, $x = y = 0$, then we obtain

$$W_{2,n}^{[3]}(\varphi;0,0) = \frac{\lambda_n}{\xi_n^2} \left(\sum_{j=1}^{2} j^2 \alpha_{j,2}^{[3]} \right) \iint_{\mathbb{D}} (s^2+t^2) e^{-(s^2+t^2)/\xi_n^2} dsdt$$

$$= \frac{\lambda_n}{\xi_n^2} \left(\alpha_{1,2}^{[3]} + 4\alpha_{2,2}^{[3]} \right) \int_{-\pi}^{\pi} \int_{0}^{\pi} \rho^3 e^{-\rho^2/\xi_n^2} d\rho d\theta$$

$$= \frac{2\pi\lambda_n}{\xi_n^2} \left(-2 + \frac{1}{2} \right) \int_{0}^{\pi} \rho^3 e^{-\rho^2/\xi_n^2} d\rho$$

$$= -\frac{3\pi\lambda_n}{\xi_n^2} \left(-\frac{\pi^2\xi_n^2 e^{-\pi^2/\xi_n^2}}{2} + \frac{\left(1 - e^{-\pi^2/\xi_n^2}\right)\xi_n^4}{2} \right)$$

$$= -\frac{3\xi_n^2}{2} + \frac{3\pi^2 e^{-\pi^2/\xi_n^2}}{2\left(1 - e^{-\pi^2/\xi_n^2}\right)} < 0,$$

by the fact that

$$1 + u \le e^u \quad \text{for all } u \ge 0.$$

We notice that the operators $W_{r,n}^{[m]}$ given by (3.4) preserve the constant functions in two variables. Indeed, for the constant function $f(x,y) = C$, by (3.1), (3.3) and (3.4), we get, for every $r,n \in \mathbb{N}$ and $m \in \mathbb{N}_0$, that

$$W_{r,n}^{[m]}(C;x,y) = \frac{C\lambda_n}{\xi_n^2} \iint_{\mathbb{D}} e^{-(s^2+t^2)/\xi_n^2} dsdt$$

$$= \frac{C\lambda_n}{\xi_n^2} \int_{-\pi}^{\pi} \int_{0}^{\pi} e^{-\rho^2/\xi_n^2} \rho d\rho d\theta$$

$$= C.$$

We also need the next lemma.

Lemma 3.1. *Let $k \in \mathbb{N}$. Then, for each $\ell = 0, 1, ..., k$ and for every $n \in \mathbb{N}$,*

$$\iint_{\mathbb{D}} s^{k-\ell} t^{\ell} e^{-(s^2+t^2)/\xi_n^2} ds dt = \begin{cases} 2\gamma_{n,k} B\left(\frac{k-\ell+1}{2}, \frac{\ell+1}{2}\right) & \text{if } k \text{ and } \ell \text{ are even} \\ 0 & \text{otherwise,} \end{cases}$$

where $B(a, b)$ denotes the Beta function, and

$$\gamma_{n,k} := \int_0^{\pi} \rho^{k+1} e^{-\rho^2/\xi_n^2} d\rho = \frac{\xi_n^{k+2}}{2} \left\{ \Gamma\left(1 + \frac{k}{2}\right) - \Gamma\left(1 + \frac{k}{2}, \left(\frac{\pi}{\xi_n}\right)^2\right) \right\},$$

(3.5)

where $\Gamma(\alpha, z) = \int_z^{\infty} t^{\alpha-1} e^{-t} dt$ is the incomplete gamma function and Γ is the gamma function.

Proof. It is clear that if k or ℓ are odd, then the integrand is a odd function with respect to s and t; and hence the above integral is zero. Also, if both k and ℓ are even, then the integrand is an even function with respect to s and t. If we define

$$\mathbb{D}_1 := \left\{ (s, t) \in \mathbb{R}^2 : 0 \leq s \leq \pi \text{ and } 0 \leq t \leq \sqrt{\pi^2 - s^2} \right\}, \qquad (3.6)$$

then we can write

$$\iint_{\mathbb{D}} s^{k-\ell} t^{\ell} e^{-(s^2+t^2)/\xi_n^2} ds dt = 4 \iint_{\mathbb{D}_1} s^{k-\ell} t^{\ell} e^{-(s^2+t^2)/\xi_n^2} ds dt$$

$$= 4 \int_0^{\pi/2} \int_0^{\pi} (\cos\theta)^{k-\ell} (\sin\theta)^{\ell} e^{-\rho^2/\xi_n^2} \rho^{k+1} d\rho d\theta$$

$$= 4\gamma_{n,k} \int_0^{\pi/2} (\cos\theta)^{k-\ell} (\sin\theta)^{\ell} d\theta$$

$$= 2\gamma_{n,k} B\left(\frac{k-\ell+1}{2}, \frac{\ell+1}{2}\right)$$

thus the result. ∎

3.2   Estimates for the Operators

Let $f \in C_{2\pi}(\mathbb{D})$, the space of all continuous functions on $\mathbb{D}$, 2π-periodic per coordinate. Then, the rth (bivariate) modulus of smoothness of f is given by (see, e.g., [33])

$$\omega_r(f; h) := \sup_{\sqrt{u^2+v^2} \leq h;\, (u,v) \in \mathbb{D}} \left\| \Delta_{u,v}^r(f) \right\| < \infty, \quad h > 0, \qquad (3.7)$$

where $\|\cdot\|$ is the sup-norm and

$$\Delta_{u,v}^r \left(f(x,y) \right) = \sum_{j=0}^{r} (-1)^{r-j} \binom{r}{j} f(x+ju, y+jv). \qquad (3.8)$$

Let $m \in \mathbb{N}_0$. By $C_{2\pi}^{(m)}(\mathbb{D})$ we mean the space of functions 2π-periodic per coordinate, having m times continuous partial derivatives with respect to the variables x and y. Observe that if $f \in C_{2\pi}^{(m)}(\mathbb{D})$, then we have that

$$\left\| \frac{\partial^m f(\cdot, \cdot)}{\partial^{m-\ell} x \partial^\ell y} \right\| := \sup_{(x,y)\in\mathbb{D}} \left| \frac{\partial^m f(x,y)}{\partial^{m-\ell} x \partial^\ell y} \right| < \infty, \qquad (3.9)$$

for every $\ell = 0, 1, ..., m$.

3.2.1 Estimates in the Case of $m \in \mathbb{N}$

Now we consider the case of $m \in \mathbb{N}$. Then, define the function

$$\begin{aligned}
G_{x,y}^{[m]}(s,t) &:= \frac{1}{(m-1)!} \sum_{j=0}^{r} \binom{r}{j} \int_0^1 (1-w)^{m-1} \\
&\times \left\{ \sum_{\ell=0}^{m} \binom{m}{m-\ell} \left| \frac{\partial^m f(x+jsw, y+jtw)}{\partial^{m-\ell} x \partial^\ell y} \right| \right\} dw
\end{aligned} \qquad (3.10)$$

for $m \in \mathbb{N}$ and $(x,y), (s,t) \in \mathbb{D}$. Notice that $G_{x,y}^{[m]}(s,t)$ is well-defined for each fixed $m \in \mathbb{N}$ when $f \in C_{2\pi}^{(m)}(\mathbb{D})$ due to the condition (3.9).

Theorem 3.2. *Let $m \in \mathbb{N}$ and $f \in C_{2\pi}^{(m)}(\mathbb{D})$. Then, for the operators $W_{r,n}^{[m]}$, we get*

$$\begin{aligned}
&\left| W_{r,n}^{[m]}(f; x, y) - f(x,y) - I_m(x,y) \right| \\
&\qquad \leq \frac{\lambda_n}{\xi_n^2} \iint_{\mathbb{D}} G_{x,y}^{[m]}(s,t) \left(|s|^m + |t|^m \right) e^{-(s^2+t^2)/\xi_n^2} ds\,dt,
\end{aligned} \qquad (3.11)$$

where λ_n is given by (3.3) and

$$\begin{aligned}
I_m(x,y) &:= \frac{2\lambda_n}{\xi_n^2} \sum_{i=1}^{[m/2]} \frac{\gamma_{n,2i} \delta_{2i,r}^{[m]}}{(2i)!} \\
&\times \left\{ \sum_{\ell=0}^{2i} B\left(\frac{2i - \ell + 1}{2}, \frac{2i + 1}{2} \right) \binom{2i}{2i - \ell} \frac{\partial^{2i} f(x,y)}{\partial^{2i-\ell} x \partial^\ell y} \right\}.
\end{aligned} \qquad (3.12)$$

The sum in (3.12) collapses when $m = 1$.

Proof. Let $(x, y) \in \mathbb{D}$ be fixed. For every $f \in C_{2\pi}(\mathbb{D})$ we can write

$$\sum_{j=0}^{r} \alpha_{j,r}^{[m]} \left(f(x + js, y + jt) - f(x, y) \right)$$

$$= \sum_{k=1}^{m} \frac{\delta_{k,r}^{[m]}}{k!} \sum_{\ell=0}^{k} \binom{k}{k-\ell} s^{k-\ell} t^{\ell} \frac{\partial^k f(x, y)}{\partial^{k-\ell} x \partial^{\ell} y}$$

$$+ \frac{1}{(m-1)!} \int_0^1 (1-w)^{m-1} \varphi_{x,y}^{[m]}(w; s, t) dw,$$

where

$$\varphi_{x,y}^{[m]}(w; s, t) := \sum_{j=0}^{r} (-1)^{r-j} \binom{r}{j}$$

$$\times \left\{ \sum_{\ell=0}^{m} \binom{m}{m-\ell} s^{m-\ell} t^{\ell} \frac{\partial^m f(x + jsw, y + jtw)}{\partial^{m-\ell} x \partial^{\ell} y} \right\}.$$

Hence, using the definition (3.4), one derives

$$W_{r,n}^{[m]}(f; x, y) - f(x, y) = \frac{\lambda_n}{\xi_n^2} \sum_{k=1}^{m} \frac{\delta_{k,r}^{[m]}}{k!} \sum_{\ell=0}^{k} \binom{k}{k-\ell} \frac{\partial^k f(x, y)}{\partial^{k-\ell} x \partial^{\ell} y}$$

$$\times \left(\iint_{\mathbb{D}} s^{k-\ell} t^{\ell} e^{-(s^2+t^2)/\xi_n^2} ds dt \right)$$

$$+ R_n^{[m]}(x, y),$$

where

$$R_n^{[m]}(x, y) := \frac{\lambda_n}{\xi_n^2 (m-1)!} \iint_{\mathbb{D}} \left(\int_0^1 (1-w)^{m-1} \varphi_{x,y}^{[m]}(w; s, t) dw \right)$$

$$\times e^{-(s^2+t^2)/\xi_n^2} ds dt.$$

Also, using Lemma 3.1, we obtain that

$$W_{r,n}^{[m]}(f; x, y) - f(x, y) - I_m(x, y) = R_n^{[m]}(x, y), \tag{3.13}$$

where $I_m(x, y)$ is given by (3.12). Because

$$\left| \varphi_{x,y}^{[m]}(w; s, t) \right| \leq (|s|^m + |t|^m) \sum_{j=0}^{r} \binom{r}{j}$$

$$\times \left\{ \sum_{\ell=0}^{m} \binom{m}{m-\ell} \left| \frac{\partial^m f(x + jsw, y + jtw)}{\partial^{m-\ell} x \partial^{\ell} y} \right| \right\},$$

it is clear that

$$\left| R_n^{[m]}(x,y) \right| \leq \frac{\lambda_n}{\xi_n^2} \iint\limits_{\mathbb{D}} G_{x,y}^{[m]}(s,t)\left(|s|^m + |t|^m \right) e^{-(s^2+t^2)/\xi_n^2} ds dt. \qquad (3.14)$$

Therefore, combining (3.13) and (3.14) the proof is finished. ∎

Corollary 3.3. *Let* $m \in \mathbb{N}$ *and* $f \in C_{2\pi}^{(m)}(\mathbb{D})$. *Then, for the operators* $W_{r,n}^{[m]}$, *we derive*

$$\left\| W_{r,n}^{[m]}(f) - f \right\| \leq \frac{C_{r,m}\lambda_n}{\xi_n^2} \left(\gamma_{n,m} + \sum_{i=1}^{[m/2]} \gamma_{n,2i} \right) \qquad (3.15)$$

for some positive constant $C_{r,m}$ *depending on* r *and* m, *where* $\gamma_{n,k}$ *is given by* (3.5). *Also, the sums in* (3.15) *collapse when* $m = 1$.

Proof. From (3.11) and (3.12), we can write

$$\left\| W_{r,n}^{[m]}(f) - f \right\| \leq \|I_m\| + \frac{\lambda_n}{\xi_n^2} \iint\limits_{\mathbb{D}} \left\| G_{x,y}^{[m]}(s,t) \right\| \left(|s|^m + |t|^m \right) e^{-(s^2+t^2)/\xi_n^2} ds dt.$$

We first estimate $\|I_m\|$. It is easy to see that

$$\|I_m\| \leq \frac{2\lambda_n}{\xi_n^2} \sum_{i=1}^{[m/2]} \frac{\gamma_{n,2i}\delta_{2i,r}^{[m]}}{(2i)!}$$

$$\times \left\{ \sum_{\ell=0}^{2i} B\left(\frac{2i-\ell+1}{2}, \frac{2i+1}{2} \right) \binom{2i}{2i-\ell} \left\| \frac{\partial^m f(\cdot,\cdot)}{\partial^{m-\ell}x \partial^\ell y} \right\| \right\}$$

$$\leq \frac{K_{r,m}\lambda_n}{\xi_n^2} \sum_{i=1}^{[m/2]} \gamma_{n,2i},$$

where

$$K_{r,m} := \max_{1 \leq i \leq [m/2]} \left\{ \frac{2\delta_{2i,r}^{[m]}}{(2i)!} \left(\sum_{\ell=0}^{2i} B\left(\frac{2i-\ell+1}{2}, \frac{2i+1}{2} \right) \binom{2i}{2i-\ell} \left\| \frac{\partial^m f(\cdot,\cdot)}{\partial^{m-\ell}x \partial^\ell y} \right\| \right) \right\}.$$

On the other hand, observe that

$$\left\| G_{x,y}^{[m]}(s,t) \right\| \leq \frac{2^r}{m!} \sum_{\ell=0}^{m} \binom{m}{m-\ell} \left\| \frac{\partial^m f(\cdot,\cdot)}{\partial^{m-\ell}x \partial^\ell y} \right\| := L_{r,m}.$$

Thus, combining these results we see that

$$\left\|W_{r,n}^{[m]}(f)-f\right\| \leq \frac{K_{r,m}\lambda_n}{\xi_n^2} \sum_{i=1}^{[m/2]} \gamma_{n,2i} + \frac{L_{r,m}\lambda_n}{\xi_n^2} \iint\limits_{\mathbb{D}} \left(|s|^m + |t|^m\right) e^{-(s^2+t^2)/\xi_n^2} ds dt$$

$$= \frac{K_{r,m}\lambda_n}{\xi_n^2} \sum_{i=1}^{[m/2]} \gamma_{n,2i} + \frac{4L_{r,m}\lambda_n}{\xi_n^2} \iint\limits_{\mathbb{D}_1} \left(s^m + t^m\right) e^{-(s^2+t^2)/\xi_n^2} ds dt$$

$$= \frac{K_{r,m}\lambda_n}{\xi_n^2} \sum_{i=1}^{[m/2]} \gamma_{n,2i}$$

$$+ \frac{4L_{r,m}\lambda_n}{\xi_n^2} \int_0^{\pi/2}\int_0^\pi \rho^{m+1}(\cos^m\theta + \sin^m\theta) e^{-\rho^2/\xi_n^2} d\rho d\theta$$

$$= \frac{K_{r,m}\lambda_n}{\xi_n^2} \sum_{i=1}^{[m/2]} \gamma_{n,2i} + \frac{4\lambda_n L_{r,m}}{\xi_n^2} B\left(\frac{m+1}{2},\frac{1}{2}\right) \gamma_{n,m},$$

which yields

$$\left\|W_{r,n}^{[m]}(f)-f\right\| \leq \frac{C_{r,m}\lambda_n}{\xi_n^2} \left(\gamma_{n,m} + \sum_{i=1}^{[m/2]} \gamma_{n,2i}\right),$$

where

$$C_{r,m} := \max\left\{K_{r,m},\ 4L_{r,m}B\left(\frac{m+1}{2},\frac{1}{2}\right)\right\}.$$

So, the proof is done. ∎

3.2.2   Estimates in the Case of $m = 0$

Now we only consider the case of $m = 0$. Then, we first obtain the following result.

Theorem 3.4. Let $f \in C_{2\pi}(\mathbb{D})$. Then, we have

$$\left|W_{r,n}^{[0]}(f;x,y) - f(x,y)\right| \leq \frac{4\lambda_n}{\xi_n^2} \iint\limits_{\mathbb{D}_1} \omega_r\left(f;\sqrt{s^2+t^2}\right) e^{-(s^2+t^2)/\xi_n^2} ds dt,$$

$$(3.16)$$

where λ_n and $\mathbb{D}_1$ are given by (3.3) and (3.6), respectively.

Proof. Let $(x,y) \in \mathbb{D}$. Taking $m = 0$ in (3.1) we see that

$$
\begin{aligned}
W_{r,n}^{[0]}(f;x,y) - f(x,y) &= \frac{\lambda_n}{\xi_n^2} \iint_{\mathbb{D}} \left\{ \sum_{j=1}^{r} \alpha_{j,r}^{[0]} \left(f\left(x+sj, y+tj\right) - f(x,y) \right) \right\} \\
&\quad \times e^{-(s^2+t^2)/\xi_n^2} ds dt \\
&= \frac{\lambda_n}{\xi_n^2} \iint_{\mathbb{D}} \left\{ \left(\sum_{j=1}^{r} (-1)^{r-j} \binom{r}{j} f\left(x+sj, y+tj\right) \right) \right. \\
&\quad \left. - \left(\sum_{j=1}^{r} (-1)^{r-j} \binom{r}{j} f(x,y) \right) \right\} e^{-(s^2+t^2)/\xi_n^2} ds dt.
\end{aligned}
$$

Then, we have

$$
\begin{aligned}
W_{r,n}^{[0]}(f;x,y) - f(x,y) &= \frac{\lambda_n}{\xi_n^2} \iint_{\mathbb{D}} \left\{ \sum_{j=0}^{r} (-1)^{r-j} \binom{r}{j} f\left(x+sj, y+tj\right) \right\} \\
&\quad \times e^{-(s^2+t^2)/\xi_n^2} ds dt
\end{aligned}
$$

and hence

$$
W_{r,n}^{[0]}(f;x,y) - f(x,y) = \frac{\lambda_n}{\xi_n^2} \iint_{\mathbb{D}} \Delta_{s,t}^{r} \left(f(x,y) \right) e^{-(s^2+t^2)/\xi_n^2} ds dt.
$$

Therefore, we find that

$$
\begin{aligned}
\left| W_{r,n}^{[0]}(f;x,y) - f(x,y) \right| &\leq \frac{\lambda_n}{\xi_n^2} \iint_{\mathbb{D}} \left| \Delta_{s,t}^{r} \left(f(x,y) \right) \right| e^{-(s^2+t^2)/\xi_n^2} ds dt \\
&\leq \frac{\lambda_n}{\xi_n^2} \iint_{\mathbb{D}} \omega_r \left(f; \sqrt{s^2+t^2} \right) e^{-(s^2+t^2)/\xi_n^2} ds dt,
\end{aligned}
$$

which completes the proof. ∎

Corollary 3.5. *Let* $f \in C_{2\pi}(\mathbb{D})$. *Then, we have*

$$
\left\| W_{r,n}^{[0]}(f) - f \right\| \leq S_r \lambda_n \omega_r (f; \xi_n) \tag{3.17}
$$

for some positive constant S_r *depending on* r.

Proof. Using (3.16) and also considering the fact that $\omega_r\left(f; \lambda u\right) \leq \left(1 + \lambda\right)^r \omega_r\left(f; u\right)$, $\lambda, u > 0$, we may write that

$$\left\| W_{r,n}^{[0]}(f) - f \right\| \leq \frac{4\lambda_n}{\xi_n^2} \iint\limits_{\mathbb{D}_1} \omega_r\left(f; \sqrt{s^2 + t^2}\right) e^{-(s^2+t^2)/\xi_n^2} ds dt$$

$$\leq \frac{4\lambda_n \omega_r\left(f; \xi_n\right)}{\xi_n^2} \iint\limits_{\mathbb{D}_1} \left(1 + \frac{\sqrt{s^2 + t^2}}{\xi_n}\right)^r e^{-(s^2+t^2)/\xi_n^2} ds dt$$

$$= \frac{4\lambda_n \omega_r\left(f; \xi_n\right)}{\xi_n^2} \int\limits_0^{\pi/2} \int\limits_0^\pi \left(1 + \frac{\rho}{\xi_n}\right)^r \rho e^{-\rho^2/\xi_n^2} d\rho d\theta$$

$$= \frac{2\pi\lambda_n \omega_r\left(f; \xi_n\right)}{\xi_n^2} \int\limits_0^\pi \left(1 + \frac{\rho}{\xi_n}\right)^r \rho e^{-\rho^2/\xi_n^2} d\rho.$$

Now setting $u = \frac{\rho}{\xi_n}$, we obtain

$$\left\| W_{r,n}^{[0]}(f) - f \right\| \leq 2\pi\lambda_n \omega_r\left(f; \xi_n\right) \int\limits_0^{\pi/\xi_n} \left(1 + u\right)^r u e^{-u^2} du$$

$$\leq 2\pi\lambda_n \omega_r\left(f; \xi_n\right) \int\limits_0^\infty \frac{\left(1 + u\right)^{r+1}}{e^{u^2}} du$$

$$=: S_r \lambda_n \omega_r\left(f; \xi_n\right)$$

where

$$S_r := 2\pi \int\limits_0^\infty \frac{\left(1 + u\right)^{r+1}}{e^{u^2}} du < \infty.$$

Therefore, the proof is finished. ∎

3.3   Statistical Approximation of the Operators

3.3.1   *Statistical Approximation in the Case of $m \in \mathbb{N}$*

We need the following lemma.

Lemma 3.6. *Let $A = [a_{jn}]$ be a non-negative regular summability matrix, and let (ξ_n) be a sequence of positive real numbers for which*

$$st_A - \lim_n \xi_n = 0. \tag{3.18}$$

Then, for each fixed $k = 1, 2, ..., m \in \mathbb{N}$, we have

$$st_A - \lim_n \frac{\gamma_{n,k}\lambda_n}{\xi_n^2} = 0,$$

where λ_n and $\gamma_{n,k}$ are given by (3.3) and (3.5), respectively.

Proof. Let $k = 1, 2, ..., m$ be fixed. Then, by (3.5), we derive

$$\frac{\gamma_{n,k}\lambda_n}{\xi_n^2} = \frac{\lambda_n}{\xi_n^2} \int_0^\pi \rho^{k+1} e^{-\rho^2/\xi_n^2} d\rho$$

$$= \frac{\lambda_n}{\xi_n^2} \int_0^\pi \rho^{k-2} \rho^2 \left(\rho e^{-\rho^2/\xi_n^2} \right) d\rho$$

$$\leq \frac{\pi^{k-2}\lambda_n}{\xi_n^2} \int_0^\pi \rho^2 \left(\rho e^{-\rho^2/\xi_n^2} \right) d\rho$$

(by change of variable and integration by parts)

$$= \frac{\pi^{k-2}\lambda_n}{\xi_n^2} \left\{ \frac{\pi^2 \xi_n^2 e^{-\pi^2/\xi_n^2}}{2} + \frac{\xi_n^4 \left(1 - e^{-\pi^2/\xi_n^2}\right)}{2} \right\}$$

Now using (3.3), we obtain that

$$\frac{\gamma_{n,k}\lambda_n}{\xi_n^2} \leq \frac{\pi^{k-1} e^{-\pi^2/\xi_n^2}}{2\left(1 - e^{-\pi^2/\xi_n^2}\right)} + \frac{\pi^{k-3} \xi_n^2}{2},$$

which gives

$$0 < \frac{\gamma_{n,k}\lambda_n}{\xi_n^2} \leq m_k \left(\frac{1}{e^{\pi^2/\xi_n^2} - 1} + \frac{\xi_n^2}{\pi^2} \right), \tag{3.19}$$

where

$$m_k := \frac{\pi^{k-1}}{2}.$$

On the other hand, the hypothesis (3.18) implies that

$$st_A - \lim_n \frac{1}{e^{\pi^2/\xi_n^2} - 1} = 0 \quad \text{and} \quad st_A - \lim_n \xi_n^2 = 0. \tag{3.20}$$

Now, for a given $\varepsilon > 0$, consider the following sets:

$$D : = \left\{ n \in \mathbb{N} : \frac{\gamma_{n,k}\lambda_n}{\xi_n^2} \geq \varepsilon \right\},$$

$$D_1 : = \left\{ n \in \mathbb{N} : \frac{1}{e^{\pi^2/\xi_n^2} - 1} \geq \frac{\varepsilon}{2m_k} \right\},$$

$$D_2 : = \left\{ n \in \mathbb{N} : \xi_n^2 \geq \frac{\varepsilon \pi^2}{2m_k} \right\}.$$

Then, from (3.19), we easily observe that

$$D \subseteq D_1 \cup D_2,$$

which yields that, for each $j \in \mathbb{N}$,

$$\sum_{j \in D} a_{jn} \leq \sum_{j \in D_1} a_{jn} + \sum_{j \in D_2} a_{jn}. \qquad (3.21)$$

Letting $j \to \infty$ in (3.21) and also using (3.20) we get

$$\lim_j \sum_{j \in D} a_{jn} = 0,$$

which completes the proof.   ∎

Now, we are ready to give the first statistical approximation theorem for the operators (3.4) in the case of $m \in \mathbb{N}$.

Theorem 3.7. *Let $A = [a_{jn}]$ be a non-negative regular summability matrix, and let (ξ_n) be a sequence of positive real numbers for which (3.18) holds. Then, for each fixed $m \in \mathbb{N}$ and for all $f \in C_{2\pi}^{(m)}(\mathbb{D})$, we have*

$$st_A - \lim_n \left\| W_{r,n}^{[m]}(f) - f \right\| = 0.$$

Proof. Let $m \in \mathbb{N}$ be fixed. Then, by (3.15), the inequality

$$\left\| W_{r,n}^{[m]}(f) - f \right\| \leq C_{r,m} \left(\frac{\gamma_{n,m}\lambda_n}{\xi_n^2} + \sum_{i=1}^{[m/2]} \frac{\gamma_{n,2i}\lambda_n}{\xi_n^2} \right) \qquad (3.22)$$

holds for some positive constant where $C_{r,m}$. Now, for a given $\varepsilon > 0$, define the following sets:

$$E := \left\{ n \in \mathbb{N} : \left\| W_{r,n}^{[m]}(f) - f \right\| \geq \varepsilon \right\},$$

$$E_i := \left\{ n \in \mathbb{N} : \frac{\gamma_{n,2i}\lambda_n}{\xi_n^2} \geq \frac{\varepsilon}{(1 + [m/2])C_{r,m}} \right\}, \ i = 1, ..., \left[\frac{m}{2} \right],$$

$$E_{1+[\frac{m}{2}]} := \left\{ n \in \mathbb{N} : \frac{\gamma_{n,m}\lambda_n}{\xi_n^2} \geq \frac{\varepsilon}{(1 + [m/2])C_{r,m}} \right\}.$$

Then, the inequality (3.22) implies that

$$E \subseteq \bigcup_{i=1}^{1+[\frac{m}{2}]} E_i,$$

and hence, for every $j \in \mathbb{N}$,

$$\sum_{n \in E} a_{jn} \leq \sum_{i=1}^{1+[\frac{m}{2}]} \sum_{n \in E_i} a_{jn}.$$

Now taking limit as $j \to \infty$ in the both sides of the above inequality and using Lemma 3.6 we get that

$$\lim_j \sum_{n \in E} a_{jn} = 0,$$

which is the desired result.   ∎

3.3.2 Statistical Approximation in the Case of $m = 0$

We now investigate the statistical approximation properties of the operators (3.4) when $m = 0$. We need the following result.

Lemma 3.8. Let $A = [a_{jn}]$ be a non-negative regular summability matrix, and let (ξ_n) be a bounded sequence of positive real numbers for which (3.18) holds. Then, for every $f \in C_{2\pi}(\mathbb{D})$, we get

$$st_A - \lim_n \lambda_n \omega_r(f; \xi_n) = 0.$$

Proof. It follows from (3.18) and (3.3) that

$$st_A - \lim_n \lambda_n = \frac{1}{\pi}.$$

Also, using the right-continuity of $\omega_r(f; \cdot)$ at zero, it is not hard to see that

$$st_A - \lim_n \omega_r(f; \xi_n) = 0.$$

Combining these results, the proof is completed.   ∎

Then, we obtain the next statistical approximation theorem.

Theorem 3.9. Let $A = [a_{jn}]$ be a non-negative regular summability matrix, and let (ξ_n) be a sequence of positive real numbers for which (3.18) holds. Then, for all $f \in C_{2\pi}(\mathbb{D})$, we get

$$st_A - \lim_n \left\| W_{r,n}^{[0]}(f) - f \right\| = 0.$$

Proof. By (3.17), the inequality

$$\left\| W_{r,n}^{[0]}(f) - f \right\| \le S_r \lambda_n \omega_r(f; \xi_n)$$

holds for some positive constant S_r. Then, for a given $\varepsilon > 0$, we can write that

$$\left\{ n \in \mathbb{N} : \left\| W_{r,n}^{[0]}(f) - f \right\| \ge \varepsilon \right\} \subseteq \left\{ n \in \mathbb{N} : \lambda_n \omega_r(f; \xi_n) \ge \frac{\varepsilon}{S_r} \right\},$$

which gives, for every $j \in \mathbb{N}$, that

$$\sum_{n:\left\| W_{r,n}^{[0]}(f)-f \right\| \geq \varepsilon} a_{jn} \leq \sum_{n:\lambda_n \omega_r(f;\xi_n) \geq \frac{\varepsilon}{8r}} a_{jn}.$$

Now, taking limit as $j \to \infty$ in the both sides of the last inequality and also using Lemma 3.8, we get that

$$\lim_j \sum_{n:\left\| W_{r,n}^{[0]}(f)-f \right\| \geq \varepsilon} a_{jn} = 0,$$

whence the result.   ∎

3.4   Conclusions

Taking $A = C_1$, the Cesáro matrix of order one, and also combining Theorems 3.7 and 3.9, we immediately obtain the following result.

Corollary 3.10. *Let* (ξ_n) *be a sequence of positive real numbers for which* $st - \lim_n \xi_n = 0$ *holds. Then, for each fixed* $m \in \mathbb{N}_0$ *and for all* $f \in C_{2\pi}^{(m)}(\mathbb{D})$, *we have* $st - \lim_n \left\| W_{r,n}^{[m]}(f) - f \right\| = 0$.

Furthermore, choosing $A = I$, the identity matrix, in Theorems 3.7 and 3.9, we have the next approximation theorems with the usual convergence.

Corollary 3.11. *Let* (ξ_n) *be a sequence of positive real numbers for which* $\lim_n \xi_n = 0$ *holds. Then, for each fixed* $m \in \mathbb{N}_0$ *and for all* $f \in C_{2\pi}^{(m)}(\mathbb{D})$, *the sequence* $\left(W_{r,n}^{[m]}(f) \right)$ *is uniformly convergent to* f *on* $\mathbb{D}$.

Now define a sequence (ξ_n) by

$$\xi_n := \begin{cases} \sqrt{n}, & \text{if } n = k^2, \ k = 1, 2, \dots \\ \frac{1}{n}, & \text{otherwise.} \end{cases} \tag{3.23}$$

Then, observe that $st - \lim_n \xi_n = 0$ although it is unbounded above. In this case, taking $A = C_1$, we obtain from Corollary 3.10 (or, Theorems 3.7 and 3.9) that

$$st - \lim_n \left\| W_{r,n}^{[m]}(f) - f \right\| = 0$$

holds for each $m \in \mathbb{N}_0$ and for all $f \in C_{2\pi}^{(m)}(\mathbb{D})$. However, since the sequence (ξ_n) given by (3.23) is non-convergent, the (classical) uniform approximation to a function f by the sequence $\left(W_{r,n}^{[m]}(f) \right)$ does not hold, i.e., Corollary 3.11 fails for the operators $W_{r,n}^{[m]}(f)$ obtained from the sequence (ξ_n) defined by (3.23).

As a result, we can say that the statistical approximation results shown in this chapter can be still valid although the operators $W_{r,n}^{[m]}$ are not positive in general and also the sequence (ξ_n) is non-convergent or unbounded.

4

Statistical L_p-Convergence of Bivariate Smooth Picard Singular Integral Operators

In this chapter, we obtain some statistical approximation results for the bivariate smooth Picard singular integral operators defined on L_p-spaces, which do not need to be positive in general. Also, giving a non-trivial example we show that the statistical L_p-approximation is stronger than the ordinary one. This chapter relies on [29].

4.1 Definition of the Operators

As usual, by $L_p\left(\mathbb{R}^2\right)$ we denote the space of all functions f defined on $\mathbb{R}^2$ for which

$$\int\limits_{-\infty}^{\infty} \int\limits_{-\infty}^{\infty} |f(x,y)|^p \, dxdy < \infty, \quad 1 \le p < \infty$$

holds. In this case, the L_p-norm of a function f in $L_p\left(\mathbb{R}^2\right)$, denoted by $\|f\|_p$, is defined to be

$$\|f\|_p = \left(\int\limits_{-\infty}^{\infty} \int\limits_{-\infty}^{\infty} |f(x,y)|^p \, dxdy \right)^{1/p}.$$

G.A. Anastassiou and O. Duman: Towards Intelligent Modeling, ISRL 14, pp. 39–60.
springerlink.com   © Springer-Verlag Berlin Heidelberg 2011

In this section, for $r \in \mathbb{N}$ and $m \in \mathbb{N}_0$, we use

$$
\alpha_{j,r}^{[m]} := \begin{cases} (-1)^{r-j} \binom{r}{j} j^{-m} & \text{if } j = 1, 2, ..., r, \\ 1 - \sum\limits_{j=1}^{r} (-1)^{r-j} \binom{r}{j} j^{-m} & \text{if } j = 0. \end{cases}
\tag{4.1}
$$

and

$$
\delta_{k,r}^{[m]} := \sum_{j=1}^{r} \alpha_{j,r}^{[m]} j^k, \quad k = 1, 2, ..., m \in \mathbb{N}.
\tag{4.2}
$$

We see that

$$
\sum_{j=0}^{r} \alpha_{j,r}^{[m]} = 1 \quad \text{and} \quad -\sum_{j=1}^{r} (-1)^{r-j} \binom{r}{j} = (-1)^r \binom{r}{0}.
\tag{4.3}
$$

Then, we consider the following bivariate smooth Picard singular integral operators:

$$
P_{r,n}^{[m]}(f; x, y) = \frac{1}{2\pi \xi_n^2} \sum_{j=0}^{r} \alpha_{j,r}^{[m]} \left(\int_{-\infty}^{\infty} \int_{-\infty}^{\infty} f(x + sj, y + tj) e^{-\sqrt{s^2 + t^2}/\xi_n} ds dt \right),
\tag{4.4}
$$

where $(x, y) \in \mathbb{R}^2$, $n, r \in \mathbb{N}$, $m \in \mathbb{N}_0$, $f \in L_p(\mathbb{R}^2)$, $1 \le p < \infty$, and also (ξ_n) is a bounded sequence of positive real numbers.

Remarks

- The operators $P_{r,n}^{[m]}$ are not in general positive. For example, take the non-negative function $\varphi(u, v) = u^2 + v^2$ and also take $r = 2$, $m = 3$, $x = 0$ and $y = 0$ in (4.4).
- It is not hard to see that the operators $P_{r,n}^{[m]}$ preserve the constant functions in two variables.
- We see, for any $\alpha > 0$, that

$$
\int_{-\infty}^{\infty} \int_{-\infty}^{\infty} e^{-\sqrt{s^2 + t^2}/\alpha} ds dt = 2\pi \alpha^2.
\tag{4.5}
$$

- Let $k \in \mathbb{N}_0$. Then, it holds, for each $\ell = 0, 1, ..., k$ and for every $n \in \mathbb{N}$, that

$$
\int_{-\infty}^{\infty} \int_{-\infty}^{\infty} s^{k-\ell} t^\ell e^{-(\sqrt{s^2 + t^2})/\xi_n} ds dt
\tag{4.6}
$$

$$
= \begin{cases} 2B\left(\frac{k-\ell+1}{2}, \frac{\ell+1}{2}\right) \xi_n^{k+2} (k+1)! & \text{if } k \text{ and } \ell \text{ are even} \\ 0 & \text{otherwise.} \end{cases}
\tag{4.7}
$$

4.2 Estimates for the Operators

As usual, let $C^{(m)}\left(\mathbb{R}^2\right)$ denote the space of all functions having m times continuous partial derivatives with respect to the variables x and y. If $f \in L_p\left(\mathbb{R}^2\right)$, then the rth (bivariate) L_p-modulus of smoothness of f is given by (see, e.g., [33])

$$\omega_r(f;h)_p := \sup_{\sqrt{u^2+v^2}\leq h} \left\|\Delta_{u,v}^r(f)\right\|_p < \infty, \quad h > 0, \ 1 \leq p < \infty, \qquad (4.8)$$

where

$$\Delta_{u,v}^r\left(f(x,y)\right) = \sum_{j=0}^{r}(-1)^{r-j}\binom{r}{j}f(x+ju,y+jv). \qquad (4.9)$$

We also use the notation

$$\partial^{r,s}f(x,y) := \frac{\partial^m f(x,y)}{\partial^r x \partial^s y} \quad \text{for } r,s = 0,1,...,m \text{ with } r+s = m.$$

We suppose that the following conditions hold:

$$f \in C^{(m)}(\mathbb{R}^2) \text{ and } \partial^{m-\ell,\ell}f(x,y) \in L_p\left(\mathbb{R}^2\right), \text{ for each } \ell = 0,1,...,m. \qquad (4.10)$$

4.2.1 Estimates in the Case of $m \in \mathbb{N}$

In this subsection, we only consider the case of $m \in \mathbb{N}$.

For $r \in \mathbb{N}$ and f satisfying (4.10), let

$$
H_{r,n}^{[m]}(x,y) : = P_{r,n}^{[m]}(f;x,y) - f(x,y)
$$
$$
-\frac{1}{2\pi\xi_n^2}\int_{-\infty}^{\infty}\int_{-\infty}^{\infty}\left(\sum_{k=1}^{m}\frac{\delta_{k,r}^{[m]}}{k!}\sum_{\ell=0}^{k}\binom{k}{k-\ell}s^{k-\ell}t^\ell\partial^{k-\ell,\ell}f(x,y)\right)dsdt.
$$

By (4.6), since, for every $r,n,m \in \mathbb{N}$,

$$
\frac{1}{2\pi\xi_n^2}\int_{-\infty}^{\infty}\int_{-\infty}^{\infty}\left(\sum_{k=1}^{m}\frac{\delta_{k,r}^{[m]}}{k!}\sum_{\ell=0}^{k}\binom{k}{k-\ell}s^{k-\ell}t^\ell\partial^{k-\ell,\ell}f(x,y)\right)dsdt
$$
$$
=\frac{1}{\pi}\sum_{i=1}^{[m/2]}(2i+1)\delta_{2i,r}^{[m]}\xi_n^{2i}\left\{\sum_{\ell=0}^{2i}\binom{2i}{2i-\ell}\partial^{2i-\ell,\ell}f(x,y)B\left(\frac{2i-\ell+1}{2},\frac{\ell+1}{2}\right)\right\},
$$

we get

$$
H_{r,n}^{[m]}(x,y) = P_{r,n}^{[m]}(f;x,y) - f(x,y) - \frac{1}{\pi}\sum_{i=1}^{[m/2]}(2i+1)\delta_{2i,r}^{[m]}\xi_n^{2i}
$$
$$
\times\left\{\sum_{\ell=0}^{2i}\binom{2i}{2i-\ell}\partial^{2i-\ell,\ell}f(x,y)B\left(\frac{2i-\ell+1}{2},\frac{\ell+1}{2}\right)\right\}.
$$
$$(4.11)$$

We also need

Lemma 4.1. *For every* $r, n, m \in \mathbb{N}$, *we get*

$$H_{r,n}^{[m]}(x,y) = \frac{1}{2\pi\xi_n^2(m-1)!} \sum_{\ell=0}^{m} \int_{-\infty}^{\infty} \int_{-\infty}^{\infty} \left(\int_0^1 (1-w)^{m-1} \Delta_{sw,tw}^r \left(\partial^{m-\ell,\ell} f(x,y) \right) dw \right)$$

$$\times \binom{m}{m-\ell} s^{m-\ell} t^{\ell} e^{-\sqrt{s^2+t^2}/\xi_n} \, ds \, dt.$$

Proof. Let $(x,y) \in \mathbb{R}^2$ be fixed. By Taylor's formula, one derives that

$$\sum_{j=0}^{r} \alpha_{j,r}^{[m]} \left(f(x+js, y+jt) - f(x,y) \right) = \sum_{k=1}^{m} \frac{\delta_{k,r}^{[m]}}{k!} \sum_{\ell=0}^{k} \binom{k}{k-\ell} s^{k-\ell} t^{\ell} \partial^{k-\ell,\ell} f(x,y)$$

$$+ \frac{1}{(m-1)!} \int_0^1 (1-w)^{m-1} \varphi_{x,y}^{[m]}(w; s, t) dw,$$

where

$$\varphi_{x,y}^{[m]}(w; s, t) := \sum_{j=0}^{r} \alpha_{j,r}^{[m]} j^m \left\{ \sum_{\ell=0}^{m} \binom{m}{m-\ell} s^{m-\ell} t^{\ell} \partial^{m-\ell,\ell} f(x+jsw, y+jtw) \right\}$$

$$- \delta_{m,r}^{[m]} \sum_{\ell=0}^{m} \binom{m}{m-\ell} s^{m-\ell} t^{\ell} \partial^{m-\ell,\ell} f(x,y).$$

We may also write that

$$\varphi_{x,y}^{[m]}(w; s, t) = \sum_{j=0}^{r} (-1)^{r-j} \binom{r}{j} \left\{ \sum_{\ell=0}^{m} \binom{m}{m-\ell} s^{m-\ell} t^{\ell} \partial^{m-\ell,\ell} f(x+jsw, y+jtw) \right\}$$

$$= \sum_{\ell=0}^{m} \binom{m}{m-\ell} s^{m-\ell} t^{\ell} \left\{ \sum_{j=0}^{r} (-1)^{r-j} \binom{r}{j} \partial^{m-\ell,\ell} f(x+jsw, y+jtw) \right\}$$

$$= \sum_{\ell=0}^{m} \binom{m}{m-\ell} s^{m-\ell} t^{\ell} \Delta_{sw,tw}^r \left(\partial^{m-\ell,\ell} f(x,y) \right).$$

Now, combining these results and also using (4.11) we obtain that

$$H_{r,n}^{[m]}(x,y) = P_{r,n}^{[m]}(f;x,y) - f(x,y) - \frac{1}{\pi} \sum_{i=1}^{[m/2]} (2i+1)\delta_{2i,r}^{[m]} \xi_n^{2i}$$

$$\times \left\{ \sum_{\ell=0}^{2i} \binom{2i}{2i-\ell} \partial^{2i-\ell,\ell} f(x,y) B\left(\frac{2i-\ell+1}{2}, \frac{\ell+1}{2}\right) \right\}$$

$$= \frac{1}{2\pi\xi_n^2(m-1)!} \sum_{\ell=0}^{m} \int_{-\infty}^{\infty} \int_{-\infty}^{\infty} \left(\int_0^1 (1-w)^{m-1} \Delta_{sw,tw}^r \left(\partial^{m-\ell,\ell} f(x,y) \right) dw \right)$$

$$\times \binom{m}{m-\ell} s^{m-\ell} t^\ell e^{-\sqrt{s^2+t^2}/\xi_n} dsdt,$$

which gives the proof.   ∎

We now present

Theorem 4.2. *Let $m,r \in \mathbb{N}$ and $p,q > 1$ such that $\frac{1}{p} + \frac{1}{q} = 1$ and f as in (4.10). Then*

$$\left\| H_{r,n}^{[m]} \right\|_p \le C\xi_n^m \left(\sum_{\ell=0}^{m} \omega_r \left(\partial^{m-\ell,\ell} f, \xi_n \right)_p^p \right)^{\frac{1}{p}},$$

for some positive constant C depending on m,p,q,r.

Proof. By Lemma 4.1, we first see that

$$\left| H_{r,n}^{[m]}(x,y) \right|^p$$

$$\le C_1 \left\{ \sum_{\ell=0}^{m} \int_{-\infty}^{\infty} \int_{-\infty}^{\infty} \left(\int_0^1 (1-w)^{m-1} \left| \Delta_{sw,tw}^r \left(\partial^{m-\ell,\ell} f(x,y) \right) \right| dw \right) \right.$$

$$\left. \times \binom{m}{m-\ell} |s|^{m-\ell} |t|^\ell e^{-\sqrt{s^2+t^2}/\xi_n} dsdt \right\}^p,$$

where

$$C_1 := \frac{1}{\left(2\pi\xi_n^2(m-1)!\right)^p}.$$

Hence, we have

$$\int_{-\infty}^{\infty} \int_{-\infty}^{\infty} \left| H_{r,n}^{[m]}(x,y) \right|^p dxdy \le C_1 \int_{-\infty}^{\infty} \int_{-\infty}^{\infty} \left(\int_{-\infty}^{\infty} \int_{-\infty}^{\infty} u_{x,y}(s,t) e^{-\sqrt{s^2+t^2}/\xi_n} dsdt \right)^p dxdy,$$

where

$$
\begin{aligned}
u_{x,y}(s,t) = \sum_{\ell=0}^{m} & \binom{m}{m-\ell} |s|^{m-\ell} |t|^{\ell} \\
& \times \left(\int_0^1 (1-w)^{m-1} \left| \Delta_{sw,tw}^r \left(\partial^{m-\ell,\ell} f(x,y) \right) \right| dw \right).
\end{aligned}
\tag{4.12}
$$

Then using Hölder's inequality for bivariate integrals and also using (4.5), we can write

$$
\int_{-\infty}^{\infty} \int_{-\infty}^{\infty} \left| H_{r,n}^{[m]}(x,y) \right|^p dx\, dy
$$

$$
\leq C_1 \int_{-\infty}^{\infty} \int_{-\infty}^{\infty} \left(\int_{-\infty}^{\infty} \int_{-\infty}^{\infty} u_{x,y}^p(s,t) e^{-\sqrt{s^2+t^2}/\xi_n} ds\, dt \right)
$$

$$
\times \left(\int_{-\infty}^{\infty} \int_{-\infty}^{\infty} e^{-(\sqrt{s^2+t^2})/\xi_n} ds\, dt \right)^{\frac{p}{q}} dx\, dy
$$

$$
:= C_2 \int_{-\infty}^{\infty} \int_{-\infty}^{\infty} \left(\int_{-\infty}^{\infty} \int_{-\infty}^{\infty} u_{x,y}^p(s,t) e^{-\sqrt{s^2+t^2}/\xi_n} ds\, dt \right) dx\, dy,
$$

where

$$
C_2 := C_1 \left(2\pi \xi_n^2 \right)^{\frac{p}{q}} = \frac{1}{2\pi \xi_n^2 \left((m-1)! \right)^p}.
$$

We now estimate $u_{x,y}^p(s,t)$. Observe that

$$
u_{x,y}(s,t)
$$

$$
\leq \sum_{\ell=0}^{m} \binom{m}{m-\ell} |s|^{m-\ell} |t|^{\ell} \left(\int_0^1 \left| \Delta_{sw,tw}^r \left(\partial^{m-\ell,\ell} f(x,y) \right) \right|^p dw \right)^{\frac{1}{p}}
$$

$$
\times \left(\int_0^1 (1-w)^{q(m-1)} dw \right)^{\frac{1}{q}}
$$

$$
:= C_3 \sum_{\ell=0}^{m} \binom{m}{m-\ell} |s|^{m-\ell} |t|^{\ell} \left(\int_0^1 \left| \Delta_{sw,tw}^r \left(\partial^{m-\ell,\ell} f(x,y) \right) \right|^p dw \right)^{\frac{1}{p}},
$$

where

$$
C_3 := \frac{1}{(q(m-1)+1)^{\frac{1}{q}}}.
$$

Hence, we get

$$u_{x,y}^p(s,t)$$

$$\leq C_3^p \left(\sum_{\ell=0}^m \binom{m}{m-\ell} |s|^{m-\ell} |t|^\ell \left(\int_0^1 |\Delta_{sw,tw}^r \left(\partial^{m-\ell,\ell} f(x,y)\right)|^p \, dw \right)^{\frac{1}{p}} \right)^p$$

$$\leq C_3^p \left\{ \sum_{\ell=0}^m \binom{m}{m-\ell} |s|^{m-\ell} |t|^\ell \left(\int_0^1 |\Delta_{sw,tw}^r \left(\partial^{m-\ell,\ell} f(x,y)\right)|^p \, dw \right) \right\}$$

$$\times \left(\sum_{\ell=0}^m \binom{m}{m-\ell} |s|^{m-\ell} |t|^\ell \right)^{\frac{p}{q}}$$

$$= C_3^p \left(|s| + |t|\right)^{\frac{mp}{q}}$$

$$\times \left\{ \sum_{\ell=0}^m \binom{m}{m-\ell} |s|^{m-\ell} |t|^\ell \left(\int_0^1 |\Delta_{sw,tw}^r \left(\partial^{m-\ell,\ell} f(x,y)\right)|^p \, dw \right) \right\}.$$

Setting

$$C_4 := C_2 C_3^p = \frac{1}{2\pi \xi_n^2 \left((m-1)!\right)^p} \frac{1}{\left(q(m-1)+1\right)^{\frac{p}{q}}}$$

and combining the above results we obtain that

$$\int_{-\infty}^{\infty} \int_{-\infty}^{\infty} \left| H_{r,n}^{[m]}(x,y) \right|^p \, dx dy$$

$$\leq C_4 \int_{-\infty}^{\infty} \int_{-\infty}^{\infty} \left\{ \int_{-\infty}^{\infty} \int_{-\infty}^{\infty} \left[\left(|s| + |t|\right)^{\frac{mp}{q}} \sum_{\ell=0}^m \binom{m}{m-\ell} |s|^{m-\ell} |t|^\ell \right. \right.$$

$$\times \left. \left. \left(\int_0^1 |\Delta_{sw,tw}^r \left(\partial^{m-\ell,\ell} f(x,y)\right)|^p \, dw \right) e^{-\sqrt{s^2+t^2}/\xi_n} \right] ds dt \right\} dx dy$$

$$= C_4 \int_{-\infty}^{\infty} \int_{-\infty}^{\infty} \left\{ \left(|s| + |t|\right)^{\frac{mp}{q}} \sum_{\ell=0}^m \binom{m}{m-\ell} |s|^{m-\ell} |t|^\ell \right.$$

$$\times \left. \left[\int_0^1 \left(\int_{-\infty}^{\infty} \int_{-\infty}^{\infty} |\Delta_{sw,tw}^r \left(\partial^{m-\ell,\ell} f(x,y)\right)|^p \, dx dy \right) dw \right] e^{-\sqrt{s^2+t^2}/\xi_n} \right\} ds dt.$$

The last gives that

$$\int\limits_{-\infty}^{\infty}\int\limits_{-\infty}^{\infty}\left|H_{r,n}^{[m]}(x,y)\right|^{p}dxdy$$

$$\leq C_{4}\int\limits_{-\infty}^{\infty}\int\limits_{-\infty}^{\infty}\left\{(|s|+|t|)^{\frac{mp}{q}}\sum_{\ell=0}^{m}\binom{m}{m-\ell}|s|^{m-\ell}|t|^{\ell}\right.$$

$$\times\left.\left[\int\limits_{0}^{1}\omega_{r}\left(\partial^{m-\ell,\ell}f,w\sqrt{s^{2}+t^{2}}\right)_{p}^{p}dw\right]e^{-\sqrt{s^{2}+t^{2}}/\xi_{n}}\right\}dsdt$$

$$=4C_{4}\int\limits_{0}^{\infty}\int\limits_{0}^{\infty}\left\{(s+t)^{\frac{mp}{q}}\sum_{\ell=0}^{m}\binom{m}{m-\ell}s^{m-\ell}t^{\ell}\right.$$

$$\times\left.\left[\int\limits_{0}^{1}\omega_{r}\left(\partial^{m-\ell,\ell}f,w\sqrt{s^{2}+t^{2}}\right)_{p}^{p}dw\right]e^{-\sqrt{s^{2}+t^{2}}/\xi_{n}}\right\}dsdt.$$

Then, considering the fact that

$$\omega_{r}\left(f,\lambda h\right)_{p}\leq(1+\lambda)^{r}\omega_{r}\left(f,h\right)_{p}\text{ for any } h,\lambda>0\text{ and }p\geq1,\qquad(4.13)$$

we have

$$\int\limits_{-\infty}^{\infty}\int\limits_{-\infty}^{\infty}\left|H_{r,n}^{[m]}(x,y)\right|^{p}dxdy$$

$$\leq4C_{4}\sum_{\ell=0}^{m}\binom{m}{m-\ell}\omega_{r}\left(\partial^{m-\ell,\ell}f,\xi_{n}\right)_{p}^{p}\int\limits_{0}^{\infty}\int\limits_{0}^{\infty}\left\{(s+t)^{\frac{mp}{q}}s^{m-\ell}t^{\ell}\right.$$

$$\times\left.\left[\int\limits_{0}^{1}\left(1+\frac{w\sqrt{s^{2}+t^{2}}}{\xi_{n}}\right)^{rp}dw\right]e^{-\sqrt{s^{2}+t^{2}}/\xi_{n}}\right\}dsdt$$

$$=C_{5}\sum_{\ell=0}^{m}\binom{m}{m-\ell}\omega_{r}\left(\partial^{m-\ell,\ell}f,\xi_{n}\right)_{p}^{p}\int\limits_{0}^{\infty}\int\limits_{0}^{\infty}\left\{(s+t)^{\frac{mp}{q}}s^{m-\ell}t^{\ell}\right.$$

$$\times\left.\left[\left(1+\frac{\sqrt{s^{2}+t^{2}}}{\xi_{n}}\right)^{rp+1}-1\right]\frac{e^{-\sqrt{s^{2}+t^{2}}/\xi_{n}}}{\sqrt{s^{2}+t^{2}}}\right\}dsdt,$$

where

$$C_{5}:=4C_{4}\left(\frac{\xi_{n}}{rp+1}\right)=\frac{2/(rp+1)}{\pi\xi_{n}\left((m-1)!\right)^{p}}\frac{1}{(q(m-1)+1)^{\frac{p}{q}}}.$$

Therefore, we see that

$$\int\limits_{-\infty}^{\infty}\int\limits_{-\infty}^{\infty}\left|H_{r,n}^{[m]}(x,y)\right|^{p}dxdy$$

$$\leq C_{5}\left\{\int\limits_{0}^{\infty}\rho^{mp}\left(\left(1+\frac{\rho}{\xi_{n}}\right)^{rp+1}-1\right)e^{-\rho/\xi_{n}}d\rho\right\}$$

$$\times\sum_{\ell=0}^{m}\binom{m}{m-\ell}\omega_{r}\left(\partial^{m-\ell,\ell}f,\xi_{n}\right)_{p}^{p}\left(\int\limits_{0}^{\pi/2}(\cos\theta+\sin\theta)^{\frac{mp}{q}}\cos^{m-\ell}\theta\sin^{\ell}\theta d\theta\right).$$

$$\leq C_{5}\left\{\int\limits_{0}^{\infty}\rho^{mp}\left(1+\frac{\rho}{\xi_{n}}\right)^{rp+1}e^{-\rho/\xi_{n}}d\rho\right\}$$

$$\times\sum_{\ell=0}^{m}\binom{m}{m-\ell}\omega_{r}\left(\partial^{m-\ell,\ell}f,\xi_{n}\right)_{p}^{p}\left(\int\limits_{0}^{\pi/2}(\cos\theta+\sin\theta)^{\frac{mp}{q}}\cos^{m-\ell}\theta\sin^{\ell}\theta d\theta\right).$$

Also, considering the fact that $0 \leq \sin\theta + \cos\theta \leq 2$ for $\theta \in [0, \frac{\pi}{2}]$, we have

$$\int\limits_{-\infty}^{\infty}\int\limits_{-\infty}^{\infty}\left|H_{r,n}^{[m]}(x,y)\right|^{p}dxdy$$

$$\leq C_{6}\left(\int\limits_{0}^{\infty}\rho^{mp}\left(1+\frac{\rho}{\xi_{n}}\right)^{rp+1}e^{-\rho/\xi_{n}}d\rho\right)\sum_{\ell=0}^{m}\binom{m}{m-\ell}\omega_{r}\left(\partial^{m,\ell}f,\xi_{n}\right)_{p}^{p}$$

$$\times\left(\int\limits_{0}^{\pi/2}\cos^{m-\ell}\theta\sin^{\ell}\theta d\theta\right),$$

where

$$C_{6}:=2^{\frac{mp}{q}}C_{5}=\frac{1}{\pi\xi_{n}\left((m-1)!\right)^{p}}\frac{1}{(q(m-1)+1)^{\frac{p}{q}}}\left(\frac{2^{\frac{mp}{q}+1}}{rp+1}\right).$$

By taking $u = \rho/\xi_n$, we obtain that

$$
\int\limits_{-\infty}^{\infty} \int\limits_{-\infty}^{\infty} \left| H_{r,n}^{[m]}(x,y) \right|^p dx dy
$$

$$
\leq \frac{C_6}{2} \xi_n^{mp+1} \left(\int\limits_0^{\infty} u^{mp} (1+u)^{rp+1} e^{-u} d\rho \right)
$$

$$
\times \left\{ \sum_{\ell=0}^{m} \binom{m}{m-\ell} B\left(\frac{m-\ell+1}{2}, \frac{\ell+1}{2} \right) \omega_r \left(\partial^{m-\ell,\ell} f, \xi_n \right)_p^p \right\}
$$

$$
\leq \frac{C_6}{2} \xi_n^{mp+1} \left(\int\limits_0^{\infty} (1+u)^{(m+r)p+1} e^{-u} d\rho \right)
$$

$$
\times \left\{ \sum_{\ell=0}^{m} \binom{m}{m-\ell} B\left(\frac{m-\ell+1}{2}, \frac{\ell+1}{2} \right) \omega_r \left(\partial^{m-\ell,\ell} f, \xi_n \right)_p^p \right\}
$$

$$
= C_7 \xi_n^{mp} \sum_{\ell=0}^{m} \binom{m}{m-\ell} B\left(\frac{m-\ell+1}{2}, \frac{\ell+1}{2} \right) \omega_r \left(\partial^{m-\ell,\ell} f, \xi_n \right)_p^p,
$$

where

$$
C_7 := \frac{1/((m-1)!)^p}{\pi \left(q(m-1)+1\right)^{\frac{p}{q}}} \left(\frac{2^{\frac{mp}{q}}}{rp+1} \right)^{(m+r)p+1} \sum_{k=0}^{(m+r)p+1} \binom{(m+r)p+1}{k} k!.
$$

Therefore the last inequality implies that

$$
\left\| H_{r,n}^{[m]} \right\|_p \leq C \xi_n^m \left(\sum_{\ell=0}^{m} \omega_r \left(\partial^{m-\ell,\ell} f, \xi_n \right)_p^p \right)^{\frac{1}{p}},
$$

where

$$
C := C(m,p,q,r)
$$

$$
= \frac{1/(m-1)!}{[\pi(rp+1)]^{\frac{1}{p}}} \left(\frac{2^m}{q(m-1)+1} \right)^{\frac{1}{q}} \left(\sum_{k=0}^{(m+r)p+1} \binom{(m+r)p+1}{k} k! \right)^{\frac{1}{p}}
$$

$$
\times \left\{ \max_{\ell=0,1,\ldots,m} \binom{m}{m-\ell} B\left(\frac{m-\ell+1}{2}, \frac{\ell+1}{2} \right) \right\}^{\frac{1}{p}}.
$$

Proof of the theorem is finished.   ∎

We also obtain the next result.

Theorem 4.3. *Let* $m, r \in \mathbb{N}$ *and* $p, q > 1$ *such that* $\frac{1}{p} + \frac{1}{q} = 1$ *and* f *is as in* (4.10) *with* $\partial^{2i-\ell,\ell} f \in L_p(\mathbb{R}^2)$, $1 \leq i \leq [m/2]$, $\ell = 0, 1, ..., 2i$. *Then*

$$\left\| P_{r,n}^{[m]}(f) - f \right\|_p \leq C\xi_n^m \left(\sum_{\ell=0}^m \omega_r \left(\partial^{m-\ell,\ell} f, \xi_n \right)_p^p \right)^{\frac{1}{p}} + B \sum_{i=1}^{[m/2]} \xi_n^{2i},$$

for some positive constants B, C *depending on* m, p, q, r.

Proof. By (4.11) and subadditivity of L_p-norm, we have

$$\left\| P_{r,n}^{[m]}(f) - f \right\|_p \leq \left\| H_{r,n}^{[m]} \right\|_p + \frac{1}{\pi} \sum_{i=1}^{[m/2]} (2i+1)\delta_{2i,r}^{[m]} \xi_n^{2i}$$

$$\times \left\{ \sum_{\ell=0}^{2i} \binom{2i}{2i-\ell} \left\| \partial^{2i-\ell,\ell} f \right\|_p B \left(\frac{2i-\ell+1}{2}, \frac{\ell+1}{2} \right) \right\}.$$

Now letting

$$B := \max_{1 \leq i \leq [m/2]} \left\{ \frac{(2i+1)\delta_{2i,r}^{[m]}}{\pi} \sum_{\ell=0}^{2i} \binom{2i}{2i-\ell} \left\| \partial^{2i-\ell,\ell} f \right\|_p B \left(\frac{2i-\ell+1}{2}, \frac{\ell+1}{2} \right) \right\},$$

by Theorem 4.2 there exists a positive constant C depending on m, p, q, r such that

$$\left\| P_{r,n}^{[m]}(f) - f \right\|_p \leq C\xi_n^m \left(\sum_{\ell=0}^m \omega_r \left(\partial^{m-\ell,\ell} f, \xi_n \right)_p^p \right)^{\frac{1}{p}} + B \sum_{i=1}^{[m/2]} \xi_n^{2i},$$

which finishes the proof. ∎

The following result gives an estimation in the cases of $p = 1$ and $m \in \mathbb{N}$.

Theorem 4.4. *Let* $m, r \in \mathbb{N}$ *and* f *as in* (4.10) *for* $p = 1$. *Then*

$$\left\| H_{r,n}^{[m]} \right\|_1 \leq D\xi_n^m \sum_{\ell=0}^m \omega_r \left(\partial^{m-\ell,\ell} f, \xi_n \right)_1$$

for some positive constant D *depending on* m, r.

Proof. By Lemma 4.1, we see that

$$\left| H_{r,n}^{[m]}(x,y) \right| \leq D_1 \sum_{\ell=0}^m \int_{-\infty}^{\infty} \int_{-\infty}^{\infty} \left(\int_0^1 (1-w)^{m-1} \left| \Delta_{sw,tw}^r \left(\partial^{m-\ell,\ell} f(x,y) \right) \right| dw \right)$$

$$\times \binom{m}{m-\ell} |s|^{m-\ell} |t|^\ell e^{-\sqrt{s^2+t^2}/\xi_n} ds dt,$$

where
$$D_1 := \frac{1}{2\pi\xi_n^2(m-1)!}.$$

Hence, we get

$$\left\|H_{r,n}^{[m]}\right\|_1 \leq D_1 \int_{-\infty}^{\infty}\int_{-\infty}^{\infty}\left\{\sum_{\ell=0}^{m}\int_{-\infty}^{\infty}\int_{-\infty}^{\infty}\left(\int_0^1 (1-w)^{m-1}\left|\Delta_{sw,tw}^r\left(\partial^{m-\ell,\ell}f(x,y)\right)\right|dw\right)\right.$$
$$\left.\times\binom{m}{m-\ell}|s|^{m-\ell}|t|^\ell e^{-\sqrt{s^2+t^2}/\xi_n}dsdt\right\}dxdy$$

$$= D_1\sum_{\ell=0}^{m}\left\{\int_{-\infty}^{\infty}\int_{-\infty}^{\infty}\left[\int_0^1(1-w)^{m-1}\left(\int_{-\infty}^{\infty}\int_{-\infty}^{\infty}\left|\Delta_{sw,tw}^r\left(\partial^{m-\ell,\ell}f(x,y)\right)\right|dxdy\right)dw\right]\right.$$
$$\left.\times\binom{m}{m-\ell}|s|^{m-\ell}|t|^\ell e^{-\sqrt{s^2+t^2}/\xi_n}dsdt\right\}.$$

Thus

$$\left\|H_{r,n}^{[m]}\right\|_1 \leq D_1\sum_{\ell=0}^{m}\left\{\int_{-\infty}^{\infty}\int_{-\infty}^{\infty}\left[\int_0^1(1-w)^{m-1}\omega_r\left(\partial^{m-\ell,\ell}f, w\sqrt{s^2+t^2}\right)_1 dw\right]\right.$$
$$\left.\times\binom{m}{m-\ell}|s|^{m-\ell}|t|^\ell e^{-(\sqrt{s^2+t^2})/\xi_n}dsdt\right\}$$

$$= 4D_1\sum_{\ell=0}^{m}\left\{\int_0^{\infty}\int_0^{\infty}\left[\int_0^1(1-w)^{m-1}\omega_r\left(\partial^{m-\ell,\ell}f, w\sqrt{s^2+t^2}\right)_1 dw\right]\right.$$
$$\left.\times\binom{m}{m-\ell}s^{m-\ell}t^\ell e^{-(\sqrt{s^2+t^2})/\xi_n}dsdt\right\}.$$

Now using (4.13) we derive that

$$\left\|H_{r,n}^{[m]}\right\|_1 \leq 4D_1\sum_{\ell=0}^{m}\binom{m}{m-\ell}\omega_r\left(\partial^{m-\ell,\ell}f, \xi_n\right)_1$$
$$\times\int_0^{\infty}\int_0^{\infty}\left(\int_0^1(1-w)^{m-1}\left(1+\frac{w\sqrt{s^2+t^2}}{\xi_n}\right)^r dw\right)s^{m-\ell}t^\ell e^{-\sqrt{s^2+t^2}/\xi_n}dsdt$$

$$\leq 4D_1\sum_{\ell=0}^{m}\binom{m}{m-\ell}\omega_r\left(\partial^{m-\ell,\ell}f, \xi_n\right)_1$$
$$\times\left\{\int_0^{\infty}\int_0^{\infty}\left(\int_0^1\left(1+\frac{w\sqrt{s^2+t^2}}{\xi_n}\right)^r dw\right)s^{m-\ell}t^\ell e^{-\sqrt{s^2+t^2}/\xi_n}dsdt\right\},$$

and so

$$\left\|H_{r,n}^{[m]}\right\|_1 \leq D_2 \sum_{\ell=0}^m \binom{m}{m-\ell} \omega_r \left(\partial^{m-\ell,\ell} f, \xi_n\right)_1$$

$$\times \left\{ \int_0^\infty \int_0^\infty \left(1 + \frac{\sqrt{s^2+t^2}}{\xi_n}\right)^{r+1} s^{m-\ell} t^\ell \frac{e^{-\sqrt{s^2+t^2}/\xi_n}}{\sqrt{s^2+t^2}} ds dt \right\},$$

where

$$D_2 = 4D_1 \frac{\xi_n}{r+1} = \frac{2}{\pi \xi_n (r+1)(m-1)!}.$$

Hence, we conclude that

$$\left\|H_{r,n}^{[m]}\right\|_1 \leq D_2 \left(\int_0^\infty \left(1 + \frac{\rho}{\xi_n}\right)^{r+1} \rho^m e^{-\rho/\xi_n} d\rho \right)$$

$$\times \sum_{\ell=0}^m \binom{m}{m-\ell} \omega_r \left(\partial^{m-\ell,\ell} f, \xi_n\right)_1 \left(\int_0^{\pi/2} \cos^{m-\ell}\theta \sin^\ell \theta d\theta \right)$$

$$= D_3 \xi_n^m \left(\int_0^\infty (1+u)^{r+1} u^m e^{-u} du \right) \sum_{\ell=0}^m \omega_r \left(\partial^{m-\ell,\ell} f, \xi_n\right)_1,$$

where

$$D_3 := \frac{1}{\pi(r+1)(m-1)!} \max_{\ell=0,1,\ldots,m} \left\{ \binom{m}{m-\ell} B\left(\frac{m-\ell+1}{2}, \frac{\ell+1}{2}\right) \right\}.$$

Also, we get that

$$\left\|H_{r,n}^{[m]}\right\|_1 \leq D_3 \xi_n^m \left(\int_0^\infty (1+u)^{m+r+1} e^{-u} du \right) \sum_{\ell=0}^m \omega_r \left(\partial^{m-\ell,\ell} f, \xi_n\right)_1$$

$$: = D \xi_n^m \sum_{\ell=0}^m \omega_r \left(\partial^{m-\ell,\ell} f, \xi_n\right)_1,$$

where

$$D := D(m,r) = \frac{\max_{\ell=0,1,\ldots,m} \left\{ \binom{m}{m-\ell} B\left(\frac{m-\ell+1}{2}, \frac{\ell+1}{2}\right) \right\}}{\pi(r+1)(m-1)!} \sum_{k=0}^{m+r+1} \binom{m+r+1}{k} k!.$$

The proof is done.   ∎

Furthermore, we obtain the following result.

Theorem 4.5. *Let $m, r \in \mathbb{N}$ and f as in (4.10) for $p = 1$ with $\partial^{2i-\ell,\ell} f \in L_1(\mathbb{R}^2)$, $1 \le i \le [m/2]$, $\ell = 0, 1, ..., 2i$. Then*

$$\left\| P_{r,n}^{[m]}(f) - f \right\|_1 \le D\xi_n^m \sum_{\ell=0}^{m} \omega_r \left(\partial^{m-\ell,\ell} f, \xi_n \right)_1 + E \sum_{i=1}^{[m/2]} \xi_n^{2i}$$

for some positive constants D, E depending on m, r.

Proof. As in the proof of Theorem 4.3, we can write

$$\left\| P_{r,n}^{[m]}(f) - f \right\|_1 \le \left\| H_{r,n}^{[m]} \right\|_1 + \frac{1}{\pi} \sum_{i=1}^{[m/2]} (2i+1)\delta_{2i,r}^{[m]} \xi_n^{2i}$$

$$\times \left\{ \sum_{\ell=0}^{2i} \binom{2i}{2i-\ell} \left\| \partial^{2i-\ell,\ell} f \right\|_1 B\left(\frac{2i-\ell+1}{2}, \frac{\ell+1}{2} \right) \right\}.$$

We put

$$E := \max_{1 \le i \le [m/2]} \left\{ \frac{(2i+1)\delta_{2i,r}^{[m]}}{\pi} \sum_{\ell=0}^{2i} \binom{2i}{2i-\ell} \left\| \partial^{2i-\ell,\ell} f \right\|_1 B\left(\frac{2i-\ell+1}{2}, \frac{\ell+1}{2} \right) \right\},$$

then we obtain from Theorem 4.4 that there exists a positive constant D depending on m, r such that

$$\left\| P_{r,n}^{[m]}(f) - f \right\|_1 \le D\xi_n^m \sum_{\ell=0}^{m} \omega_r \left(\partial^{m-\ell,\ell} f, \xi_n \right)_1 + E \sum_{i=1}^{[m/2]} \xi_n^{2i},$$

whence the proof. ∎

4.2.2   Estimates in the Case of $m = 0$

We now focus on the estimation in the case of $m = 0$. We first obtain the next result.

Theorem 4.6. *Let $r \in \mathbb{N}$ and $p, q > 1$ such that $\frac{1}{p} + \frac{1}{q} = 1$. Then, for every $f \in L_p(\mathbb{R}^2)$, the following inequality*

$$\left\| P_{r,n}^{[0]}(f) - f \right\|_p \le K\omega_r \left(f, \xi_n \right)_p$$

holds for some positive constant K depending on p, r.

Proof. By (4.1), (4.3) and (4.4), we can write

$$P_{r,n}^{[0]}(f;x,y) - f(x,y)$$

$$= \frac{1}{2\pi\xi_n^2} \int\limits_{-\infty}^{\infty} \int\limits_{-\infty}^{\infty} \left\{ \sum_{j=1}^{r} (-1)^{r-j} \binom{r}{j} \left(f(x+sj, y+tj) - f(x,y) \right) \right\}$$
$$\times e^{-\sqrt{s^2+t^2}/\xi_n} ds dt$$

$$= \frac{1}{2\pi\xi_n^2} \int\limits_{-\infty}^{\infty} \int\limits_{-\infty}^{\infty} \sum_{j=1}^{r} \left\{ (-1)^{r-j} \binom{r}{j} f(x+sj, y+tj) + (-1)^{r} \binom{r}{0} f(x,y) \right\}$$
$$\times e^{-\sqrt{s^2+t^2}/\xi_n} ds dt$$

$$= \frac{1}{2\pi\xi_n^2} \int\limits_{-\infty}^{\infty} \int\limits_{-\infty}^{\infty} \left\{ \sum_{j=0}^{r} (-1)^{r-j} \binom{r}{j} f(x+sj, y+tj) \right\} e^{-\sqrt{s^2+t^2}/\xi_n} ds dt.$$

Also, by (4.9), we have

$$P_{r,n}^{[0]}(f;x,y) - f(x,y) = \frac{1}{2\pi\xi_n^2} \int\limits_{-\infty}^{\infty} \int\limits_{-\infty}^{\infty} \Delta_{s,t}^r (f(x,y)) \, e^{-(\sqrt{s^2+t^2})/\xi_n} ds dt,$$

which yields

$$\left| P_{r,n}^{[0]}(f;x,y) - f(x,y) \right| \leq \frac{1}{2\pi\xi_n^2} \int\limits_{-\infty}^{\infty} \int\limits_{-\infty}^{\infty} \left| \Delta_{s,t}^r (f(x,y)) \right| e^{-\sqrt{s^2+t^2}/\xi_n} ds dt.$$

$$(4.14)$$

Hence, we obtain that

$$\int\limits_{-\infty}^{\infty} \int\limits_{-\infty}^{\infty} \left| P_{r,n}^{[0]}(f;x,y) - f(x,y) \right|^p dx dy$$

$$\leq K_1 \int\limits_{-\infty}^{\infty} \int\limits_{-\infty}^{\infty} \left(\int\limits_{-\infty}^{\infty} \int\limits_{-\infty}^{\infty} \left| \Delta_{s,t}^r (f(x,y)) \right| e^{-\sqrt{s^2+t^2}/\xi_n} ds dt \right)^p dx dy$$

where

$$K_1 := \frac{1}{\left(2\pi\xi_n^2 \right)^p}.$$

Now, we get from Hölder's inequality for bivariate integrals that

$$
\int\limits_{-\infty}^{\infty} \int\limits_{-\infty}^{\infty} \left| P_{r,n}^{[0]}(f;x,y) - f(x,y) \right|^p dxdy
$$

$$
\leq K_1 \int\limits_{-\infty}^{\infty} \int\limits_{-\infty}^{\infty} \left\{ \left(\int\limits_{-\infty}^{\infty} \int\limits_{-\infty}^{\infty} \left| \Delta_{s,t}^r \left(f(x,y) \right) \right|^p e^{-\sqrt{s^2+t^2}/\xi_n} dsdt \right) \right.
$$

$$
\left. \times \left(\int\limits_{-\infty}^{\infty} \int\limits_{-\infty}^{\infty} e^{-\sqrt{s^2+t^2})/\xi_n} dsdt \right)^{p/q} \right\} dxdy.
$$

Then, using (4.5), we can write

$$
\int\limits_{-\infty}^{\infty} \int\limits_{-\infty}^{\infty} \left| P_{r,n}^{[0]}(f;x,y) - f(x,y) \right|^p dxdy
$$

$$
\leq K_1 \left(2\pi \xi_n^2 \right)^{p/q} \int\limits_{-\infty}^{\infty} \int\limits_{-\infty}^{\infty} \left(\int\limits_{-\infty}^{\infty} \int\limits_{-\infty}^{\infty} \left| \Delta_{s,t}^r \left(f(x,y) \right) \right|^p e^{-\sqrt{s^2+t^2}/\xi_n} dsdt \right) dxdy
$$

$$
:= K_2 \int\limits_{-\infty}^{\infty} \int\limits_{-\infty}^{\infty} \left(\int\limits_{-\infty}^{\infty} \int\limits_{-\infty}^{\infty} \left| \Delta_{s,t}^r \left(f(x,y) \right) \right|^p e^{-(p\sqrt{s^2+t^2}/2\xi_n} dsdt \right) dxdy,
$$

where

$$
K_2 := K_1 \left(2\pi \xi_n^2 \right)^{p/q} = \frac{1}{2\pi \xi_n^2}.
$$

Thus, by (4.13), we get

$$
\int\limits_{-\infty}^{\infty} \int\limits_{-\infty}^{\infty} \left| P_{r,n}^{[0]}(f;x,y) - f(x,y) \right|^p dxdy
$$

$$
\leq K_2 \int\limits_{-\infty}^{\infty} \int\limits_{-\infty}^{\infty} \omega_r \left(f, \sqrt{s^2+t^2} \right)_p^p e^{-\sqrt{s^2+t^2}/\xi_n} dsdt
$$

$$
= 4K_2 \int\limits_{0}^{\infty} \int\limits_{0}^{\infty} \omega_r \left(f, \sqrt{s^2+t^2} \right)_p^p e^{-\sqrt{s^2+t^2}/\xi_n} dsdt
$$

$$
\leq 4K_2 \omega_r \left(f, \xi_n \right)_p^p \int\limits_{0}^{\infty} \int\limits_{0}^{\infty} \left(1 + \frac{\sqrt{s^2+t^2}}{\xi_n} \right)^{rp} e^{-\sqrt{s^2+t^2}/\xi_n} dsdt.
$$

After some calculations, we derive that

$$\int\limits_{-\infty}^{\infty} \int\limits_{-\infty}^{\infty} \left| P_{r,n}^{[0]}(f;x,y) - f(x,y) \right|^p dxdy$$

$$\leq 4K_2\omega_r\left(f,\xi_n\right)_p^p \int\limits_0^{\pi/2} \int\limits_0^{\infty} \left(1 + \frac{\rho}{\xi_n}\right)^{rp} e^{-\rho/\xi_n} \rho d\rho d\theta$$

$$= \omega_r\left(f,\xi_n\right)_p^p \int\limits_0^{\infty} (1+u)^{rp}\, e^{-u} u\, du$$

$$\leq \omega_r\left(f,\xi_n\right)_p^p \int\limits_0^{\infty} (1+u)^{rp+1}\, e^{-u} du$$

$$= \left(\sum\limits_{k=0}^{r+1} \binom{r+1}{k} k!\right) \omega_r\left(f,\xi_n\right)_p^p .$$

Therefore, we have

$$\left\| P_{r,n}^{[0]}(f) - f \right\|_p \leq K\omega_r\left(f,\xi_n\right)_p ,$$

where

$$K := K(p,r) = \left(\sum\limits_{k=0}^{r+1} \binom{r+1}{k} k!\right)^{\frac{1}{p}} .$$

The proof of the theorem is done.   ∎

Finally we obtain an estimation in the case of $p = 1$ and $m = 0$.

Theorem 4.7. *For every* $f \in L_1\left(\mathbb{R}^2\right)$, *we get*

$$\left\| P_{r,n}^{[0]}(f) - f \right\|_1 \leq L\omega_r\left(f,\xi_n\right)_1$$

for some positive constant L *depending on* r.

Proof. By (4.14), we easily see that

$$\left\| P_{r,n}^{[0]}(f) - f \right\|_1 = \int_{-\infty}^{\infty} \int_{-\infty}^{\infty} \left| P_{r,n}^{[0]}(f; x, y) - f(x, y) \right| dx dy$$

$$\leq \frac{1}{2\pi\xi_n^2} \int_{-\infty}^{\infty} \int_{-\infty}^{\infty} \left(\int_{-\infty}^{\infty} \int_{-\infty}^{\infty} \left| \Delta_{s,t}^r (f(x,y)) \right| e^{-\sqrt{s^2+t^2}/\xi_n} ds dt \right) dx dy$$

$$= \frac{1}{2\pi\xi_n^2} \int_{-\infty}^{\infty} \int_{-\infty}^{\infty} \left(\int_{-\infty}^{\infty} \int_{-\infty}^{\infty} \left| \Delta_{s,t}^r (f(x,y)) \right| dx dy \right) e^{-\sqrt{s^2+t^2}/\xi_n} ds dt$$

$$\leq \frac{1}{2\pi\xi_n^2} \int_{-\infty}^{\infty} \int_{-\infty}^{\infty} \omega_r \left(f, \sqrt{s^2 + t^2} \right)_1 e^{-(\sqrt{s^2+t^2})/\xi_n} ds dt$$

$$= \frac{2}{\pi\xi_n^2} \int_0^{\infty} \int_0^{\infty} \omega_r \left(f, \sqrt{s^2 + t^2} \right)_1 e^{-(\sqrt{s^2+t^2})/\xi_n} ds dt.$$

Now using (4.13), we have

$$\left\| P_{r,n}^{[0]}(f) - f \right\|_1 \leq \frac{2\omega_r (f, \xi_n)_1}{\pi\xi_n^2} \int_0^{\infty} \int_0^{\infty} \left(1 + \frac{\sqrt{s^2 + t^2}}{\xi_n} \right)^r e^{-\sqrt{s^2+t^2}/\xi_n} ds dt$$

$$= \frac{2\omega_r (f, \xi_n)_1}{\pi\xi_n^2} \int_0^{\pi/2} \int_0^{\infty} \left(1 + \frac{\rho}{\xi_n} \right)^r e^{-\rho/\xi_n} \rho d\rho d\theta$$

$$= \omega_r (f, \xi_n)_1 \int_0^{\infty} (1 + u)^r e^{-u} u \, du$$

$$\leq \omega_r (f, \xi_n)_1 \int_0^{\infty} (1 + u)^{r+1} e^{-u} du.$$

Then, the last inequality implies that

$$\left\| P_{r,n}^{[0]}(f) - f \right\|_1 \leq L\omega_r (f, \xi_n)_1,$$

where

$$L := L(r) = \sum_{k=0}^{r+1} \binom{r+1}{k} k!.$$

The proof is done.   ∎

4.3 Statistical L_p-Approximation of the Operators

In order to get the statistical approximation properties of the operators (4.4) we first need the following lemma.

Lemma 4.8. *Let $A = [a_{jn}]$ be a non-negative regular summability matrix, and let (ξ_n) be a bounded sequence of positive real numbers for which*

$$st_A - \lim_n \xi_n = 0 \qquad (4.15)$$

holds. Then, for every $f \in L_p(\mathbb{R}^2)$ with $1 \leq p < \infty$, we get

$$st_A - \lim_n \omega_r(f; \xi_n)_p = 0. \qquad (4.16)$$

Proof. Let $1 \leq p < \infty$. By the right-continuity of $\omega_r(f; \cdot)_p$ at zero, we can write that, for a given $\varepsilon > 0$, there exists a $\delta > 0$ such that $\omega_r(f; h)_p < \varepsilon$ whenever $0 < h < \delta$. Hence, $\omega_r(f; h)_p \geq \varepsilon$ yields that $h \geq \delta$. Now replacing h by ξ_n, for every $\varepsilon > 0$, we observe that

$$\{n : \omega_r(f; \xi_n)_p \geq \varepsilon\} \subseteq \{n : \xi_n \geq \delta\},$$

which gives that, for each $j \in \mathbb{N}$,

$$\sum_{n : \omega_r(f;\xi_n)_p \geq \varepsilon} a_{jn} \leq \sum_{n : \xi_n \geq \delta} a_{jn}.$$

Also, by (4.15), we have

$$\lim_j \sum_{n : \xi_n \geq \delta} a_{jn} = 0.$$

The last equality gives that

$$\lim_j \sum_{n : \omega_r(f;\xi_n)_p \geq \varepsilon} a_{jn} = 0,$$

which implies (4.16). So, the proof is finished. ∎

4.3.1 Statistical L_p-Approximation in the Case of $m \in \mathbb{N}$

Combining Theorems 4.3 and 4.5 we immediately obtain the following result.

Corollary 4.9. *Let $1 \leq p < \infty$ and $m \in \mathbb{N}$. Then, for every f as in (4.10) with $\partial^{2i-\ell,\ell} f \in L_p(\mathbb{R}^2)$, $1 \leq i \leq [m/2]$, $\ell = 0, 1, ..., 2i$, we get*

$$\left\| P_{r,n}^{[m]}(f) - f \right\|_p \leq M_1 \left\{ \sum_{\ell=0}^m \left(\xi_n^m \omega_r \left(\partial^{m-\ell,\ell} f, \xi_n \right)_p \right)^p \right\}^{\frac{1}{p}} + M_2 \sum_{i=1}^{[m/2]} \xi_n^{2i}$$

for some positive constants M_1, M_2 depending on m, p, q, r, where

$$M_1 := \begin{cases} D \ (\text{as in Theorem 4.5}) \ \text{if } p = 1 \\ C \ (\text{as in Theorem 4.3}) \ \text{if } 1 < p < \infty \ \text{with } (1/p) + (1/q) = 1 \end{cases}$$

and

$$M_2 := \begin{cases} E \ (\text{as in Theorem 4.5}) \ \text{if } p = 1 \\ B \ (\text{as in Theorem 4.3}) \ \text{if } 1 < p < \infty \ \text{with } (1/p) + (1/q) = 1. \end{cases}$$

Now we are ready to give the following statistical L_p-approximation result.

Theorem 4.10. *Let $m, r \in \mathbb{N}$ and $A = [a_{jn}]$ be a non-negative regular summability matrix, and let (ξ_n) be a bounded sequence of positive real numbers for which (4.15) holds. Then, for all f as in (4.10) with $\partial^{2i-\ell,\ell} f \in L_p(\mathbb{R}^2)$, $1 \leq i \leq [m/2]$, $\ell = 0, 1, ..., 2i$; $1 \leq p < \infty$, we get*

$$st_A - \lim_n \|P_{r,n}(f) - f\|_p = 0. \tag{4.17}$$

Proof. From (4.15) and Lemma 4.8 we can write

$$st_A - \lim_n \left(\xi_n^m \omega_r \left(\partial^{m-\ell,\ell} f, \xi_n \right)_p \right)^p = 0 \ \text{for each } \ell = 0, 1, ..., m \tag{4.18}$$

and

$$st_A - \lim_n \xi_n^{2i} = 0 \ \text{for each } i = 1, 2, ..., \left[\frac{m}{2}\right]. \tag{4.19}$$

Now, for a given $\varepsilon > 0$, consider the following sets:

$$S := \left\{ n \in \mathbb{N} : \left\| P_{r,n}^{[m]}(f) - f \right\|_p \geq \varepsilon \right\},$$

$$S_\ell := \left\{ n \in \mathbb{N} : \left(\xi_n^m \omega_r \left(\partial^{m-\ell,\ell} f, \xi_n \right)_p \right)^p \geq \frac{\varepsilon}{(m + [m/2] + 1) M_1} \right\},$$

$$(\ell = 0, 1, ..., m),$$

$$S_{i+m} := \left\{ n \in \mathbb{N} : \xi_n^{2i} \geq \frac{\varepsilon}{(m + [m/2] + 1) M_2} \right\} \ (i = 1, 2, ..., \left[\frac{m}{2}\right]).$$

Thus, by Corollary 4.9 we have

$$S \subseteq \bigcup_{k=0}^{m+[m/2]} S_k,$$

which gives, for every $j \in \mathbb{N}$, that

$$\sum_{n \in S} a_{jn} \leq \sum_{k=0}^{m+[m/2]} \sum_{n \in S_k} a_{jn}.$$

Now, taking limit as $j \to \infty$ in the both sides of the last inequality and also using (4.18), (4.19), we derive that

$$\lim_{j} \sum_{n \in S} a_{jn} = 0,$$

which implies (4.17). Hence, the proof is done.   ∎

4.3.2 Statistical L_p-Approximation in the Case of $m = 0$

In this subsection, we first combining Theorems 4.6 and 4.7 as follows:

Corollary 4.11. *Let $1 \le p < \infty$ and $r \in \mathbb{N}$. Then, for every $f \in L_p(\mathbb{R}^2)$, we get*

$$\left\| P_{r,n}^{[0]}(f) - f \right\|_p \le N \omega_r (f, \xi_n)_p$$

for some positive constant N depending on p, r, where

$$N := \begin{cases} L \ (\text{as in Theorem 4.7}) & \text{if } p = 1 \\ K \ (\text{as in Theorem 4.6}) & \text{if } 1 < p < \infty \text{ with } (1/p) + (1/q) = 1. \end{cases}$$

Now we can give the second statistical L_p-approximation result.

Theorem 4.12. *Let $r \in \mathbb{N}$ and $A = [a_{jn}]$ be a non-negative regular summability matrix, and let (ξ_n) be a bounded sequence of positive real numbers for which (4.15) holds. Then, for all $f \in L_p(\mathbb{R}^2)$ with $1 \le p < \infty$, we get*

$$st_A - \lim_{n} \left\| P_{r,n}^{[0]}(f) - f \right\|_p = 0. \tag{4.20}$$

Proof. It follows from Corollary 4.11 that, for every $\varepsilon > 0$,

$$\left\{ n \in \mathbb{N} : \left\| P_{r,n}^{[0]}(f) - f \right\|_p \ge \varepsilon \right\} \subseteq \left\{ n \in \mathbb{N} : \omega_r (f, \xi_n)_p \ge \frac{\varepsilon}{N} \right\}.$$

Hence, for each $j \in \mathbb{N}$, we get

$$\sum_{n: \left\| P_{r,n}^{[0]}(f) - f \right\|_p \ge \varepsilon} a_{jn} \le \sum_{n: \omega_r (f, \xi_n)_p \ge \frac{\varepsilon}{N}} a_{jn}.$$

Now, letting $j \to \infty$ in the last inequality and also considering Lemma 4.8, we see that

$$\lim_{j} \sum_{n: \left\| P_{r,n}^{[0]}(f) - f \right\|_p \ge \varepsilon} a_{jn} = 0,$$

which gives (4.20).   ∎

4.4 Conclusions

In this section, we give some special cases of the approximation results obtained in this chapter.

In particular, we first consider the case of $A = C_1$, the Cesáro matrix of order one. In this case, from Theorems 4.10 and 4.12 we get the next result at once.

Corollary 4.13. *Let $m \in \mathbb{N}_0$, $r \in \mathbb{N}$, and let (ξ_n) be a bounded sequence of positive real numbers for which $st - \lim_n \xi_n = 0$ holds. Then, for all f as in (4.10) with $\partial^{2i-\ell,\ell} f \in L_p(\mathbb{R}^2)$, $1 \leq i \leq [m/2]$, $\ell = 0, 1, ..., 2i$; $1 \leq p < \infty$, we get*

$$st - \lim_n \left\| P_{r,n}^{[m]}(f) - f \right\|_p = 0.$$

The second result is the case of $A = I$, the identity matrix. Then, the next approximation theorem is a direct consequence of Theorems 4.10 and 4.12.

Corollary 4.14. *Let $m \in \mathbb{N}_0$, $r \in \mathbb{N}$, and let (ξ_n) be a bounded sequence of positive real numbers for which $\lim_n \xi_n = 0$ holds. Then, for all f as in (4.10) with $\partial^{2i-\ell,\ell} f \in L_p(\mathbb{R}^2)$, $1 \leq i \leq [m/2]$, $\ell = 0, 1, ..., 2i$; $1 \leq p < \infty$, the sequence $\left(P_{r,n}^{[m]}(f) \right)$ is uniformly convergent to f with respect to the L_p-norm.*

Finally, define a sequence (ξ_n) as follows:

$$\xi_n := \begin{cases} \frac{n}{1+n}, & \text{if } n = k^2, \ k = 1, 2, ... \\ \frac{n}{1+n^2}, & \text{otherwise.} \end{cases} \tag{4.21}$$

Then, we see that $st - \lim_n \xi_n = 0$. So, if we use this sequence (ξ_n) in the definition of the operator $P_{r,n}^{[m]}$, then, we derive from Corollary 4.13 (or, Theorems 4.10 and 4.12) that $st - \lim_n \left\| P_{r,n}^{[m]}(f) - f \right\|_p = 0$ holds for all f as in (4.10) with $\partial^{2i-\ell,\ell} f \in L_p(\mathbb{R}^2)$, $1 \leq i \leq [m/2]$, $\ell = 0, 1, ..., 2i$; $1 \leq p < \infty$. However, because the sequence (ξ_n) given by (4.21) is non-convergent, the classical L_p-approximation to a function f by the operators $P_{r,n}^{[m]}(f)$ is impossible, i.e., Corollary 4.14 fails for these operators. We should note that Theorems 4.10 and 4.12, and Corollary 4.13 are also valid when $\lim \xi_n = 0$ since every convergent sequence is A-statistically convergent, and so statistically convergent. But, as in the above example, the theorems obtained in this chapter still work although (ξ_n) is non-convergent. Therefore, this non-trivial example clearly shows that the statistical L_p-approximation results in Theorems 4.10 and 4.12, and also in Corollary 4.13 are more applicable than Corollary 4.14.

5

Statistical L_p-Approximation by Bivariate Gauss-Weierstrass Singular Integral Operators

In this chapter, we study statistical L_p-approximation properties of the bivariate Gauss-Weierstrass singular integral operators which are not positive in general. Furthermore, we introduce a non-trivial example showing that the statistical L_p-approximation is more powerful than the ordinary case. This chapter relies on [23].

5.1 Definition of the Operators

Consider the set $\mathbb{D}$ given by

$$\mathbb{D} := \left\{ (s,t) \in \mathbb{R}^2 : s^2 + t^2 \leq \pi^2 \right\}.$$

As usual, by $L_p(\mathbb{D})$ we denote the space of all functions f defined on $\mathbb{D}$ for which

$$\iint\limits_{\mathbb{D}} |f(x,y)|^p \, dxdy < \infty, \quad 1 \leq p < \infty$$

holds. In this case, the L_p-norm of a function f in $L_p(\mathbb{D})$, denoted by $\|f\|_p$, is given by

$$\|f\|_p = \left(\iint\limits_{\mathbb{D}} |f(x,y)|^p \, dxdy \right)^{1/p}.$$

G.A. Anastassiou and O. Duman: Towards Intelligent Modeling, ISRL 14, pp. 61–83.
springerlink.com

For $r \in \mathbb{N}$ and $m \in \mathbb{N}_0 := \mathbb{N} \cup \{0\}$, we use

$$\alpha_{j,r}^{[m]} := \begin{cases} (-1)^{r-j} \binom{r}{j} j^{-m} & \text{if } j = 1, 2, ..., r, \\ 1 - \sum_{j=1}^{r} (-1)^{r-j} \binom{r}{j} j^{-m} & \text{if } j = 0. \end{cases} \tag{5.1}$$

and

$$\delta_{k,r}^{[m]} := \sum_{j=1}^{r} \alpha_{j,r}^{[m]} j^k, \quad k = 1, 2, ..., m \in \mathbb{N}. \tag{5.2}$$

We see that

$$\sum_{j=0}^{r} \alpha_{j,r} = 1 \quad \text{and} \quad -\sum_{j=1}^{r} (-1)^{r-j} \binom{r}{j} = (-1)^r \binom{r}{0}. \tag{5.3}$$

Suppose that (ξ_n) is a sequence of positive real numbers. Letting

$$\lambda_n := \frac{1}{\pi \left(1 - e^{-\pi^2/\xi_n^2}\right)} \quad \left(\lambda_n \to \frac{1}{\pi}, \text{ as } \xi_n \to 0\right), \tag{5.4}$$

we define the following bivariate smooth Gauss-Weierstrass singular integral operators:

$$W_{r,n}^{[m]}(f; x, y) = \frac{\lambda_n}{\xi_n^2} \sum_{j=0}^{r} \alpha_{j,r}^{[m]} \left(\iint_{\mathbb{D}} f(x + sj, y + tj) e^{-(s^2+t^2)/\xi_n^2} ds dt \right), \tag{5.5}$$

where $(x, y) \in \mathbb{D}$, $n, r \in \mathbb{N}$, $m \in \mathbb{N}_0$ and $f \in L_p(\mathbb{D})$, $1 \le p < \infty$.

Remarks.

- The operators $W_{r,n}^{[m]}$ are not in general positive. For example, consider the non-negative function $\varphi(u, v) = u^2 + v^2$ and also take $r = 2$, $m = 3$, $x = 0$ and $y = 0$ in (5.5).
- It is easy to check that the operators $W_{r,n}^{[m]}$ preserve the constant functions in two variables.
- We obtain, for any $\alpha > 0$, that

$$\iint_{\mathbb{D}} e^{-(s^2+t^2)/\alpha} ds dt = \alpha \pi \left(1 - e^{-\pi^2/\alpha}\right). \tag{5.6}$$

- Let $k \in \mathbb{N}_0$. Then, it holds, for each $\ell = 0, 1, ..., k$ and for every $n \in \mathbb{N}$, that

$$\iint_{\mathbb{D}} s^{k-\ell} t^\ell e^{-(s^2+t^2)/\xi_n^2} ds dt = \begin{cases} 2\gamma_{n,k} B\left(\frac{k-\ell+1}{2}, \frac{\ell+1}{2}\right) & \text{if } k \text{ and } \ell \text{ are even} \\ 0 & \text{otherwise,} \end{cases}$$

$$\tag{5.7}$$

where $B(a, b)$ denotes the Beta function, and

$$\gamma_{n,k} := \int_0^\pi \rho^{k+1} e^{-\rho^2/\xi_n^2} d\rho = \frac{\xi_n^{k+2}}{2} \left\{ \Gamma\left(1 + \frac{k}{2}\right) - \Gamma\left(1 + \frac{k}{2}, \left(\frac{\pi}{\xi_n}\right)^2\right) \right\},$$

(5.8)

where $\Gamma(\alpha, z) = \int_z^\infty t^{\alpha-1} e^{-t} dt$ is the incomplete gamma function and Γ is the gamma function.

5.2 Estimates for the Operators

For $f \in L_p(\mathbb{D})$ and 2π-periodic per coordinate, the rth (bivariate) L_p-modulus of smoothness of f is given by (see, e.g., [33])

$$\omega_r(f; h)_p := \sup_{\sqrt{u^2+v^2} \leq h} \left\| \Delta_{u,v}^r(f) \right\|_p < \infty, \quad h > 0, \ 1 \leq p < \infty,$$

(5.9)

where

$$\Delta_{u,v}^r(f(x,y)) = \sum_{j=0}^r (-1)^{r-j} \binom{r}{j} f(x + ju, y + jv).$$

(5.10)

We also use the notation

$$\partial^{m-\ell, \ell} f(x, y) := \frac{\partial^m f(x, y)}{\partial^{m-\ell} x \partial^\ell y} \quad \text{for} \ \ell = 0, 1, ..., m.$$

Assume that

$$f \in C_{2\pi}^{(m)}(\mathbb{D}),$$

(5.11)

the space of functions 2π-periodic per coordinate, having m times continuous partial derivatives with respect to the variables x and y, $m \in \mathbb{N}_0$.

5.2.1 Estimates in the Case of $m \in \mathbb{N}$

In this subsection, we only consider the case of $m \in \mathbb{N}$.

For $r \in \mathbb{N}$ and f satisfying (5.11), let

$$H_{r,n}^{[m]}(x, y) := W_{r,n}^{[m]}(f; x, y) - f(x, y)$$

$$- \frac{\lambda_n}{\xi_n^2} \iint_{\mathbb{D}} \left(\sum_{k=1}^m \frac{\delta_{k,r}^{[m]}}{k!} \sum_{\ell=0}^k \binom{k}{k-\ell} s^{k-\ell} t^\ell \partial^{k-\ell, \ell} f(x, y) \right) e^{-(s^2+t^2)/\xi_n^2} ds dt.$$

By (5.7), since, for every $r, n, m \in \mathbb{N}$,

$$\iint_{\mathbb{D}} \left(\sum_{k=1}^m \frac{\delta_{k,r}^{[m]}}{k!} \sum_{\ell=0}^k \binom{k}{k-\ell} s^{k-\ell} t^\ell \partial^{k-\ell, \ell} f(x, y) \right) e^{-(s^2+t^2)/\xi_n^2} ds dt$$

$$= 2 \sum_{i=1}^{[m/2]} \frac{\delta_{2i,r}^{[m]} \gamma_{n,2i}}{(2i)!} \left\{ \sum_{\ell=0}^{2i} \binom{2i}{2i-\ell} \partial^{2i-\ell, \ell} f(x, y) B\left(\frac{2i-\ell+1}{2}, \frac{\ell+1}{2}\right) \right\},$$

where $[\cdot]$ is the integral part, we get

$$H_{r,n}^{[m]}(x,y) = W_{r,n}^{[m]}(f;x,y) - f(x,y) - \frac{2\lambda_n}{\xi_n^2} \sum_{i=1}^{[m/2]} \frac{\delta_{2i,r}^{[m]} \gamma_{n,2i}}{(2i)!}$$
$$\times \left\{ \sum_{\ell=0}^{2i} \binom{2i}{2i-\ell} \partial^{2i-\ell,\ell} f(x,y) B \left(\frac{2i-\ell+1}{2}, \frac{\ell+1}{2} \right) \right\},$$

$$(5.12)$$

where $\gamma_{n,k}$ is given by (5.8). Now we obtain the next result.

Lemma 5.1. *For every $r,n,m \in \mathbb{N}$ and for all f satisfying (5.11), we get*

$$H_{r,n}^{[m]}(x,y) = \frac{\lambda_n}{\xi_n^2(m-1)!} \sum_{\ell=0}^{m} \iint_{\mathbb{D}} \left(\int_0^1 (1-w)^{m-1} \Delta_{sw,tw}^r \left(\partial^{m-\ell,\ell} f(x,y) \right) dw \right)$$
$$\times \binom{m}{m-\ell} s^{m-\ell} t^\ell e^{-(s^2+t^2)/\xi_n^2} ds dt.$$

Proof. Let $(x,y) \in \mathbb{D}$ be fixed. By Taylor's formula, one can write

$$\sum_{j=0}^{r} \alpha_{j,r}^{[m]} \left(f(x+js, y+jt) - f(x,y) \right) = \sum_{k=1}^{m} \frac{\delta_{k,r}^{[m]}}{k!} \sum_{\ell=0}^{k} \binom{k}{k-\ell} s^{k-\ell} t^\ell \partial^{k-\ell,\ell} f(x,y)$$
$$+ \frac{1}{(m-1)!} \int_0^1 (1-w)^{m-1} \varphi_{x,y}^{[m]}(w;s,t) dw,$$

where

$$\varphi_{x,y}^{[m]}(w;s,t) := \sum_{j=0}^{r} \alpha_{j,r}^{[m]} j^m \left\{ \sum_{\ell=0}^{m} \binom{m}{m-\ell} s^{m-\ell} t^\ell \partial^{m-\ell,\ell} f(x+jsw, y+jtw) \right\}$$
$$- \delta_{m,r}^{[m]} \sum_{\ell=0}^{m} \binom{m}{m-\ell} s^{m-\ell} t^\ell \partial^{m-\ell,\ell} f(x,y)$$
$$= \sum_{\ell=0}^{m} \binom{m}{m-\ell} s^{m-\ell} t^\ell \Delta_{sw,tw}^r \left(\partial^{m-\ell,\ell} f(x,y) \right).$$

Then, by (5.12) we have

$$H_{r,n}^{[m]}(x,y) = W_{r,n}^{[m]}(f;x,y) - f(x,y) - \frac{2\lambda_n}{\xi_n^2} \sum_{i=1}^{[m/2]} \frac{\delta_{2i,r}^{[m]}\gamma_{n,2i}}{(2i)!}$$

$$\times \left\{ \sum_{\ell=0}^{2i} \binom{2i}{2i-\ell} \partial^{2i-\ell,\ell} f(x,y) B\left(\frac{2i-\ell+1}{2}, \frac{\ell+1}{2}\right) \right\}$$

$$= \frac{\lambda_n}{\xi_n^2(m-1)!} \sum_{\ell=0}^{m} \iint_{\mathbb{D}} \left(\int_0^1 (1-w)^{m-1} \Delta_{sw,tw}^r \left(\partial^{m-\ell,\ell}f(x,y)\right) dw \right)$$

$$\times \binom{m}{m-\ell} s^{m-\ell} t^\ell e^{-(s^2+t^2)/\xi_n^2} ds dt,$$

which finishes the proof. ∎

Theorem 5.2. *Let $m,r \in \mathbb{N}$ and $p,q > 1$ such that $\frac{1}{p} + \frac{1}{q} = 1$ and $f \in C_\pi^{(m)}(\mathbb{D})$. Then the following inequality*

$$\left\| H_{r,n}^{[m]} \right\|_p \leq \frac{C\xi_n^m}{(1-e^{-\pi^2/\xi_n^2})^{\frac{1}{p}}} \left(\sum_{\ell=0}^{m} \omega_r \left(\partial^{m-\ell,\ell}f, \xi_n\right)_p^p \right)^{\frac{1}{p}}$$

holds for some positive constant C depending on m,p,q,r.

Proof. By Lemma 5.1, we first observe that

$$\left| H_{r,n}^{[m]}(x,y) \right|^p$$

$$\leq \frac{\lambda_n^p}{\xi_n^{2p}((m-1)!)^p} \left\{ \sum_{\ell=0}^{m} \iint_{\mathbb{D}} \left(\int_0^1 (1-w)^{m-1} \left| \Delta_{sw,tw}^r \left(\partial^{m-\ell,\ell}f(x,y)\right) \right| dw \right) \right.$$

$$\left. \times \binom{m}{m-\ell} |s|^{m-\ell} |t|^\ell e^{-(s^2+t^2)/\xi_n^2} ds dt \right\}^p$$

$$= \frac{C_1}{\xi_n^{2p} (1-e^{-\pi^2/\xi_n^2})^p} \left\{ \sum_{\ell=0}^{m} \iint_{\mathbb{D}} \left(\int_0^1 (1-w)^{m-1} \left| \Delta_{sw,tw}^r \left(\partial^{m-\ell,\ell}f(x,y)\right) \right| dw \right) \right.$$

$$\left. \times \binom{m}{m-\ell} |s|^{m-\ell} |t|^\ell e^{-(s^2+t^2)/\xi_n^2} ds dt \right\}^p$$

where

$$C_1 := \frac{1}{\pi^p((m-1)!)^p}.$$

Hence, we obtain that

$$\iint\limits_{\mathbb{D}} \left| H_{r,n}^{[m]}(x,y) \right|^p dxdy$$

$$\leq \frac{C_1}{\xi_n^{2p} \left(1 - e^{-\pi^2/\xi_n^2}\right)^p} \iint\limits_{\mathbb{D}} \left(\iint\limits_{\mathbb{D}} u_{x,y}(s,t) e^{-(s^2+t^2)/\xi_n^2} dsdt \right)^p dxdy.$$

where

$$
\begin{aligned}
u_{x,y}(s,t) = \sum_{\ell=0}^{m} & \binom{m}{m-\ell} |s|^{m-\ell} |t|^{\ell} \\
& \times \left(\int_0^1 (1-w)^{m-1} \left| \Delta_{sw,tw}^r \left(\partial^{m-\ell,\ell} f(x,y) \right) \right| dw \right).
\end{aligned}
\tag{5.13}
$$

Then using the Hölder's inequality for bivariate integrals and also considering (5.6), we can write

$$\iint\limits_{\mathbb{D}} \left| H_{r,n}^{[m]}(x,y) \right|^p dxdy$$

$$\leq \frac{C_1}{\xi_n^{2p} \left(1 - e^{-\pi^2/\xi_n^2}\right)^p} \iint\limits_{\mathbb{D}} \left(\iint\limits_{\mathbb{D}} u_{x,y}^p(s,t) e^{-(s^2+t^2)/\xi_n^2} dsdt \right) dxdy$$

$$\times \left(\iint\limits_{\mathbb{D}} e^{-(s^2+t^2)/\xi_n^2} dsdt \right)^{\frac{p}{q}}$$

$$= C_1 \frac{\left\{ \pi \xi_n^2 \left(1 - e^{-\pi^2/\xi_n^2}\right) \right\}^{\frac{p}{q}}}{\xi_n^{2p} \left(1 - e^{-\pi^2/\xi_n^2}\right)^p} \iint\limits_{\mathbb{D}} \left(\iint\limits_{\mathbb{D}} u_{x,y}^p(s,t) e^{-(s^2+t^2)/\xi_n^2} dsdt \right) dxdy$$

$$:= \frac{C_2}{\xi_n^2 \left(1 - e^{-\pi^2/\xi_n^2}\right)} \iint\limits_{\mathbb{D}} \left(\iint\limits_{\mathbb{D}} u_{x,y}^p(s,t) e^{-(s^2+t^2)/\xi_n^2} dsdt \right) dxdy,$$

where

$$C_2 := C_1 \pi^{\frac{p}{q}} = \frac{1}{\pi \left((m-1)!\right)^p}.$$

We now estimate $u_{x,y}^p(s,t)$. We see that

$$u_{x,y}(s,t) \leq \sum_{\ell=0}^{m} \binom{m}{m-\ell} |s|^{m-\ell} |t|^{\ell} \left(\int_0^1 \left| \Delta_{sw,tw}^r \left(\partial^{m-\ell,\ell} f(x,y) \right) \right|^p dw \right)^{\frac{1}{p}}$$

$$\times \left(\int_0^1 (1-w)^{q(m-1)} dw \right)^{\frac{1}{q}}$$

$$:= C_3 \sum_{\ell=0}^{m} \binom{m}{m-\ell} |s|^{m-\ell} |t|^{\ell} \left(\int_0^1 \left| \Delta_{sw,tw}^r \left(\partial^{m-\ell,\ell} f(x,y) \right) \right|^p dw \right)^{\frac{1}{p}},$$

where

$$C_3 := \frac{1}{(q(m-1)+1)^{\frac{1}{q}}}.$$

Hence, we get

$$u_{x,y}^p(s,t) \leq C_3^p \left(\sum_{\ell=0}^{m} \binom{m}{m-\ell} |s|^{m-\ell} |t|^{\ell} \left(\int_0^1 \left| \Delta_{sw,tw}^r \left(\partial^{m-\ell,\ell} f(x,y) \right) \right|^p dw \right)^{\frac{1}{p}} \right)^p,$$

which implies

$$u_{x,y}^p(s,t) \leq C_3^p \left\{ \sum_{\ell=0}^{m} \binom{m}{m-\ell} |s|^{m-\ell} |t|^{\ell} \left(\int_0^1 \left| \Delta_{sw,tw}^r \left(\partial^{m-\ell,\ell} f(x,y) \right) \right|^p dw \right) \right\}$$

$$\times \left(\sum_{\ell=0}^{m} \binom{m}{m-\ell} |s|^{m-\ell} |t|^{\ell} \right)^{\frac{p}{q}}$$

$$= C_3^p \left(|s| + |t| \right)^{\frac{mp}{q}}$$

$$\times \left\{ \sum_{\ell=0}^{m} \binom{m}{m-\ell} |s|^{m-\ell} |t|^{\ell} \left(\int_0^1 \left| \Delta_{sw,tw}^r \left(\partial^{m-\ell,\ell} f(x,y) \right) \right|^p dw \right) \right\}.$$

Setting

$$C_4 := C_2 C_3^p = \frac{1/\left((m-1)!\right)^p}{\pi \left(q(m-1)+1 \right)^{\frac{p}{q}}}$$

and combining the above results we obtain

$$\iint\limits_{\mathbb{D}} \left| H_{r,n}^{[m]}(x,y) \right|^p dxdy$$

$$\leq \frac{C_4}{\xi_n^2 \left(1 - e^{-\pi^2/\xi_n^2}\right)} \iint\limits_{\mathbb{D}} \left\{ \iint\limits_{\mathbb{D}} \left[(|s| + |t|)^{\frac{mp}{q}} \sum_{\ell=0}^m \binom{m}{m-\ell} |s|^{m-\ell} |t|^\ell \right. \right.$$

$$\times \left. \left. \left(\int\limits_0^1 \left| \Delta_{sw,tw}^r \left(\partial^{m-\ell,\ell} f(x,y) \right) \right|^p dw \right) e^{-(s^2+t^2)/\xi_n^2} \right] dsdt \right\} dxdy$$

$$= \frac{C_4}{\xi_n^2 \left(1 - e^{-\pi^2/\xi_n^2}\right)} \iint\limits_{\mathbb{D}} \left\{ (|s| + |t|)^{\frac{mp}{q}} \sum_{\ell=0}^m \binom{m}{m-\ell} |s|^{m-\ell} |t|^\ell \right.$$

$$\times \left. \left[\int\limits_0^1 \left(\iint\limits_{\mathbb{D}} \left| \Delta_{sw,tw}^r \left(\partial^{m-\ell,\ell} f(x,y) \right) \right|^p dxdy \right) dw \right] e^{-(s^2+t^2)/\xi_n^2} \right\} dsdt,$$

which yields that

$$\iint\limits_{\mathbb{D}} \left| H_{r,n}^{[m]}(x,y) \right|^p dxdy$$

$$\leq \frac{C_4}{\xi_n^2 \left(1 - e^{-\pi^2/\xi_n^2}\right)} \iint\limits_{\mathbb{D}} \left\{ (|s| + |t|)^{\frac{mp}{q}} \sum_{\ell=0}^m \binom{m}{m-\ell} |s|^{m-\ell} |t|^\ell \right.$$

$$\times \left. \left[\int\limits_0^1 \omega_r \left(\partial^{m-\ell,\ell} f, w\sqrt{s^2 + t^2} \right)_p^p dw \right] e^{-(s^2+t^2)/\xi_n^2} \right\} dsdt.$$

Thus, we get

$$\iint\limits_{\mathbb{D}} \left| H_{r,n}^{[m]}(x,y) \right|^p dxdy$$

$$\leq \frac{4C_4}{\xi_n^2 \left(1 - e^{-\pi^2/\xi_n^2}\right)} \iint\limits_{\mathbb{D}_1} \left\{ (s + t)^{\frac{mp}{q}} \sum_{\ell=0}^m \binom{m}{m-\ell} s^{m-\ell} t^\ell \right.$$

$$\times \left. \left[\int\limits_0^1 \omega_r \left(\partial^{m-\ell,\ell} f, w\sqrt{s^2 + t^2} \right)_p^p dw \right] e^{-(s^2+t^2)/\xi_n^2} \right\} dsdt,$$

where

$$\mathbb{D}_1 := \left\{ (s,t) \in \mathbb{R}^2 : 0 \leq s \leq \pi \text{ and } 0 \leq t \leq \sqrt{\pi^2 - s^2} \right\}. \qquad (5.14)$$

Now, using the fact that

$$\omega_r \left(f, \lambda h\right)_p \le (1 + \lambda)^r \, \omega_r \left(f, h\right)_p \ \text{for any } h, \lambda > 0 \text{ and } p \ge 1, \qquad (5.15)$$

we have

$$\iint_{\mathbb{D}} \left| H_{r,n}^{[m]}(x, y) \right|^p dx dy$$

$$\le \frac{4C_4}{\xi_n^2 \left(1 - e^{-\pi^2/\xi_n^2}\right)} \sum_{\ell=0}^{m} \binom{m}{m - \ell} \omega_r \left(\partial^{m-\ell, \ell} f, \xi_n\right)_p^p \iint_{\mathbb{D}_1} \left\{ (s + t)^{\frac{mp}{q}} s^{m-\ell} t^{\ell} \right.$$

$$\times \left[\int_0^1 \left(1 + \frac{w\sqrt{s^2 + t^2}}{\xi_n}\right)^{rp} dw \right] e^{-(s^2+t^2)/\xi_n^2} \bigg\} ds dt,$$

and hence

$$\iint_{\mathbb{D}} \left| H_{r,n}^{[m]}(x, y) \right|^p dx dy$$

$$\le \frac{C_5}{\xi_n \left(1 - e^{-\pi^2/\xi_n^2}\right)} \sum_{\ell=0}^{m} \binom{m}{m - \ell} \omega_r \left(\partial^{m-\ell, \ell} f, \xi_n\right)_p^p \iint_{\mathbb{D}_1} \left\{ (s + t)^{\frac{mp}{q}} s^{m-\ell} t^{\ell} \right.$$

$$\times \left[\left(1 + \frac{\sqrt{s^2 + t^2}}{\xi_n}\right)^{rp+1} - 1 \right] \frac{e^{-(s^2+t^2)/\xi_n^2}}{\sqrt{s^2 + t^2}} \bigg\} ds dt,$$

where

$$C_5 := \frac{4C_4}{rp + 1} = \left(\frac{4}{rp + 1}\right) \frac{1/\left((m - 1)!\right)^p}{\pi \left(q(m - 1) + 1\right)^{\frac{p}{q}}}.$$

Therefore, we derive that

$$\iint\limits_{\mathbb{D}} \left| H_{r,n}^{[m]}(x,y) \right|^p dxdy$$

$$\leq \frac{C_5}{\xi_n \left(1 - e^{-\pi^2/\xi_n^2}\right)} \left\{ \int_0^\pi \rho^{mp} \left(\left(1 + \frac{\rho}{\xi_n}\right)^{rp+1} - 1 \right) e^{-\rho^2/\xi_n^2} d\rho \right\}$$

$$\times \sum_{\ell=0}^m \binom{m}{m-\ell} \omega_r \left(\partial^{m-\ell,\ell}f, \xi_n\right)_p^p \left(\int_0^{\pi/2} (\cos\theta + \sin\theta)^{\frac{mp}{q}} \cos^{m-\ell}\theta \sin^\ell \theta d\theta \right)$$

$$\leq \frac{C_5}{\xi_n \left(1 - e^{-\pi^2/\xi_n^2}\right)} \left\{ \int_0^\pi \rho^{mp} \left(1 + \frac{\rho}{\xi_n}\right)^{rp+1} e^{-\rho^2/\xi_n^2} d\rho \right\}$$

$$\times \sum_{\ell=0}^m \binom{m}{m-\ell} \omega_r \left(\partial^{m-\ell,\ell}f, \xi_n\right)_p^p \left(\int_0^{\pi/2} (\cos\theta + \sin\theta)^{\frac{mp}{q}} \cos^{m-\ell}\theta \sin^\ell \theta d\theta \right).$$

Using the fact that $0 \leq \sin\theta + \cos\theta \leq 2$ for $\theta \in [0, \frac{\pi}{2}]$, we have

$$\iint\limits_{\mathbb{D}} \left| H_{r,n}^{[m]}(x,y) \right|^p dxdy$$

$$\leq \frac{C_6}{\xi_n \left(1 - e^{-\pi^2/\xi_n^2}\right)} \left(\int_0^\pi \rho^{mp} \left(1 + \frac{\rho}{\xi_n}\right)^{rp+1} e^{-\rho^2/\xi_n^2} d\rho \right)$$

$$\times \sum_{\ell=0}^m \binom{m}{m-\ell} \omega_r \left(\partial^{m-\ell,\ell}f, \xi_n\right)_p^p \left(\int_0^{\pi/2} \cos^{m-\ell}\theta \sin^\ell \theta d\theta \right),$$

where

$$C_6 := 2^{\frac{mp}{q}} C_5 = \left(\frac{2^{\frac{mp}{q}+2}}{rp+1} \right) \frac{1/((m-1)!)^p}{\pi \left(q(m-1)+1\right)^{\frac{p}{q}}}.$$

If we take $u = \rho/\xi_n$, then we observe that

$$\iint\limits_{\mathbb{D}} \left| H_{r,n}^{[m]}(x,y) \right|^p dxdy$$

$$\leq \frac{C_6 \xi_n^{mp}}{2 \left(1 - e^{-\pi^2/\xi_n^2}\right)} \left(\int_0^{\pi/\xi_n} u^{mp} (1+u)^{rp+1} e^{-u^2} du \right)$$

$$\times \left\{ \sum_{\ell=0}^m \binom{m}{m-\ell} B\left(\frac{m-\ell+1}{2}, \frac{\ell+1}{2} \right) \omega_r \left(\partial^{m-\ell,\ell}f, \xi_n\right)_p^p \right\},$$

which gives

$$\iint\limits_{\mathbb{D}} \left| H_{r,n}^{[m]}(x,y) \right|^p dxdy$$

$$\leq \frac{C_6 \xi_n^{mp}}{2\left(1 - e^{-\pi^2/\xi_n^2}\right)} \left(\int\limits_0^\infty (1+u)^{(m+r)p+1} e^{-u^2} du \right)$$

$$\times \left\{ \sum_{\ell=0}^m \binom{m}{m-\ell} B\left(\frac{m-\ell+1}{2}, \frac{\ell+1}{2}\right) \omega_r \left(\partial^{m-\ell,\ell} f, \xi_n\right)_p^p \right\}$$

$$= \frac{C_7 \xi_n^{mp}}{\left(1 - e^{-\pi^2/\xi_n^2}\right)} \sum_{\ell=0}^m \binom{m}{m-\ell} B\left(\frac{m-\ell+1}{2}, \frac{\ell+1}{2}\right) \omega_r \left(\partial^{m-\ell,\ell} f, \xi_n\right)_p^p,$$

where

$$C_7 := \left(\frac{2^{\frac{mp}{q}+1}}{rp+1} \right) \frac{1/\left((m-1)!\right)^p}{\pi \left(q(m-1)+1\right)^{\frac{p}{q}}} \left(\int\limits_0^\infty (1+u)^{(m+r)p+1} e^{-u^2} du \right).$$

Therefore the last inequality implies that

$$\left\| H_{r,n}^{[m]} \right\|_p \leq \frac{C\xi_n^m}{\left(1 - e^{-\pi^2/\xi_n^2}\right)^{\frac{1}{p}}} \left(\sum_{\ell=0}^m \omega_r \left(\partial^{m-\ell,\ell} f, \xi_n\right)_p^p \right)^{\frac{1}{p}},$$

where

$$C := C(m,p,q,r)$$

$$= \frac{1/(m-1)!}{\pi^{\frac{1}{p}} \left(q(m-1)+1\right)^{\frac{1}{q}}} \left(\frac{2^{\frac{mp}{q}+1}}{rp+1} \right)^{\frac{1}{p}} \left(\int\limits_0^\infty (1+u)^{(m+r)p+1} e^{-u^2} du \right)^{\frac{1}{p}}$$

$$\times \left\{ \max_{\ell=0,1,\ldots,m} \binom{m}{m-\ell} B\left(\frac{m-\ell+1}{2}, \frac{\ell+1}{2}\right) \right\}^{\frac{1}{p}}.$$

The proof is done.   ∎

We also obtain the following result.

Theorem 5.3. *Let $m,r \in \mathbb{N}$ and $p,q > 1$ such that $\frac{1}{p} + \frac{1}{q} = 1$ and $f \in C_{2\pi}^{(m)}(\mathbb{D})$. Then, the inequality*

$$\left\| W_{r,n}^{[m]}(f) - f \right\|_p \leq \frac{C\xi_n^m}{\left(1 - e^{-\pi^2/\xi_n^2}\right)^{\frac{1}{p}}} \left(\sum_{\ell=0}^m \omega_r \left(\partial^{m-\ell,\ell} f, \xi_n\right)_p^p \right)^{\frac{1}{p}} + B\lambda_n \sum_{i=1}^{[m/2]} \xi_n^{2i}$$

holds for some positive constants B, C depending on m, p, q, r; B also depends on f.

Proof. By (5.12) and subadditivity of L_p-norm, we have

$$
\left\| W_{r,n}^{[m]}(f) - f \right\|_p \leq \left\| H_{r,n}^{[m]} \right\|_p + \frac{2\lambda_n}{\xi_n^2} \sum_{i=1}^{[m/2]} \frac{\delta_{2i,r}^{[m]} \gamma_{n,2i}}{(2i)!}
$$

$$
\times \left\{ \sum_{\ell=0}^{2i} \binom{2i}{2i-\ell} \left\| \partial^{2i-\ell,\ell} f \right\|_p B\left(\frac{2i-\ell+1}{2}, \frac{\ell+1}{2} \right) \right\}
$$

$$
\leq \left\| H_{r,n}^{[m]} \right\|_p + \lambda_n \sum_{i=1}^{[m/2]} \frac{\delta_{2i,r}^{[m]} \xi_n^{2i}}{(i+1)...(2i)}
$$

$$
\times \left\{ \sum_{\ell=0}^{2i} \binom{2i}{2i-\ell} \left\| \partial^{2i-\ell,\ell} f \right\|_p B\left(\frac{2i-\ell+1}{2}, \frac{\ell+1}{2} \right) \right\}.
$$

Now by setting

$$
B := \max_{1 \leq i \leq [m/2]} \left\{ \frac{\delta_{2i,r}^{[m]}}{(i+1)...(2i)} \sum_{\ell=0}^{2i} \binom{2i}{2i-\ell} \left\| \partial^{2i-\ell,\ell} f \right\|_p B\left(\frac{2i-\ell+1}{2}, \frac{\ell+1}{2} \right) \right\},
$$

we obtain

$$
\left\| W_{r,n}^{[m]}(f) - f \right\|_p \leq \left\| H_{r,n}^{[m]} \right\|_p + B\lambda_n \sum_{i=1}^{[m/2]} \xi_n^{2i},
$$

by Theorem 5.2, now claim is proved.   ∎

The following result gives an estimation in the cases of $p = 1$ and $m \in \mathbb{N}$.

Theorem 5.4. *Let $m, r \in \mathbb{N}$ and $f \in C_\pi^{(m)}(\mathbb{D})$. Then, we get*

$$
\left\| H_{r,n}^{[m]} \right\|_1 \leq \frac{D\xi_n^m}{(1 - e^{-\pi^2/\xi_n^2})} \sum_{\ell=0}^m \omega_r \left(\partial^{m-\ell,\ell} f, \xi_n \right)_1
$$

for some positive constant D depending on m, r.

Proof. By Lemma 5.1, we see that

$$
\left| H_{r,n}^{[m]}(x,y) \right| \leq \frac{\lambda_n}{\xi_n^2 (m-1)!} \sum_{\ell=0}^m \iint_{\mathbb{D}} \left(\int_0^1 (1-w)^{m-1} \left| \Delta_{sw,tw}^r \left(\partial^{m-\ell,\ell} f(x,y) \right) \right| dw \right.
$$

$$
\times \binom{m}{m-\ell} |s|^{m-\ell} |t|^\ell e^{-(s^2+t^2)/\xi_n^2} \, ds dt.
$$

Then, we have

$$
\left\| H_{r,n}^{[m]} \right\|_1 \leq \frac{\lambda_n}{\xi_n^2(m-1)!} \iint\limits_{\mathbb{D}} \left\{ \sum_{\ell=0}^m \iint\limits_{\mathbb{D}} \left(\int_0^1 (1-w)^{m-1} \left| \Delta_{sw,tw}^r \left(\partial^{m-\ell,\ell} f(x,y) \right) \right| dw \right) \right.
$$
$$
\times \binom{m}{m-\ell} |s|^{m-\ell} |t|^\ell \, e^{-(s^2+t^2)/\xi_n^2} ds dt \Bigg\} dx dy
$$
$$
= \frac{\lambda_n}{\xi_n^2(m-1)!} \sum_{\ell=0}^m \left\{ \iint\limits_{\mathbb{D}} \left[\int_0^1 (1-w)^{m-1} \left(\iint\limits_{\mathbb{D}} \left| \Delta_{sw,tw}^r \left(\partial^{m-\ell,\ell} f(x,y) \right) \right| dx dy \right) dw \right] \right.
$$
$$
\times \binom{m}{m-\ell} |s|^{m-\ell} |t|^\ell \, e^{-(s^2+t^2)/\xi_n^2} ds dt \Bigg\},
$$

which implies that

$$
\left\| H_{r,n}^{[m]} \right\|_1 \leq \frac{\lambda_n}{\xi_n^2(m-1)!} \sum_{\ell=0}^m \left\{ \iint\limits_{\mathbb{D}} \left[\int_0^1 (1-w)^{m-1} \omega_r \left(\partial^{m-\ell,\ell} f, w\sqrt{s^2+t^2} \right)_1 dw \right] \right.
$$
$$
\times \binom{m}{m-\ell} |s|^{m-\ell} |t|^\ell \, e^{-(s^2+t^2)/\xi_n^2} ds dt \Bigg\}
$$
$$
= \frac{4\lambda_n}{\xi_n^2(m-1)!} \sum_{\ell=0}^m \left\{ \iint\limits_{\mathbb{D}_1} \left[\int_0^1 (1-w)^{m-1} \omega_r \left(\partial^{m-\ell,\ell} f, w\sqrt{s^2+t^2} \right)_1 dw \right] \right.
$$
$$
\times \binom{m}{m-\ell} s^{m-\ell} t^\ell e^{-(s^2+t^2)/\xi_n^2} ds dt \Bigg\},
$$

where the set $\mathbb{D}_1$ is given by (5.14). Now using (5.15) we derive that

$$
\left\| H_{r,n}^{[m]} \right\|_1 \leq \frac{4\lambda_n}{\xi_n^2(m-1)!} \sum_{\ell=0}^m \binom{m}{m-\ell} \omega_r \left(\partial^{m-\ell,\ell} f, \xi_n \right)_1
$$
$$
\times \iint\limits_{\mathbb{D}_1} \left(\int_0^1 (1-w)^{m-1} \left(1 + \frac{w\sqrt{s^2+t^2}}{\xi_n} \right)^r dw \right) s^{m-\ell} t^\ell e^{-(s^2+t^2)/\xi_n^2} ds dt
$$
$$
\leq \frac{4\lambda_n}{\xi_n^2(m-1)!} \sum_{\ell=0}^m \binom{m}{m-\ell} \omega_r \left(\partial^{m-\ell,\ell} f, \xi_n \right)_1
$$
$$
\times \left\{ \iint\limits_{\mathbb{D}_1} \left(\int_0^1 \left(1 + \frac{w\sqrt{s^2+t^2}}{\xi_n} \right)^r dw \right) s^{m-\ell} t^\ell e^{-(s^2+t^2)/\xi_n^2} ds dt \right\},
$$

and hence

$$\left\|H_{r,n}^{[m]}\right\|_1 \leq \frac{D'}{\xi_n\left(1-e^{-\pi^2/\xi_n^2}\right)} \sum_{\ell=0}^{m}\binom{m}{m-\ell}\omega_r\left(\partial^{m-\ell,\ell}f,\xi_n\right)_1$$
$$\times\left\{\iint\limits_{\mathbb{D}_1}\left(1+\frac{\sqrt{s^2+t^2}}{\xi_n}\right)^{r+1}s^{m-\ell}t^\ell\frac{e^{-(s^2+t^2)/\xi_n^2}}{\sqrt{s^2+t^2}}dsdt\right\},$$

where

$$D'=\frac{4}{\pi\left(r+1\right)\left(m-1\right)!}.$$

Thus, we obtain that

$$\left\|H_{r,n}^{[m]}\right\|_1 \leq \frac{D'}{\xi_n\left(1-e^{-\pi^2/\xi_n^2}\right)}\left(\int_0^\pi\left(1+\frac{\rho}{\xi_n}\right)^{r+1}\rho^m e^{-\rho^2/\xi_n^2}d\rho\right)$$
$$\times\sum_{\ell=0}^{m}\binom{m}{m-\ell}\omega_r\left(\partial^{m-\ell,\ell}f,\xi_n\right)_1\left(\int_0^{\pi/2}\cos^{m-\ell}\theta\sin^\ell\theta d\theta\right)$$
$$=\frac{D'\xi_n^m}{\left(1-e^{-\pi^2/\xi_n^2}\right)}\left(\int_0^{\pi/\xi_n}\left(1+u\right)^{r+1}u^m e^{-u^2}du\right)$$
$$\times\sum_{\ell=0}^{m}\binom{m}{m-\ell}\omega_r\left(\partial^{m-\ell,\ell}f,\xi_n\right)_1 B\left(\frac{m-\ell+1}{2},\frac{\ell+1}{2}\right).$$

Now taking

$$D:=D'\left(\int_0^\infty\left(1+u\right)^{m+r+1}u^m e^{-u^2}du\right)\max_{\ell=0,1,\ldots,m}\left\{\binom{m}{m-\ell}B\left(\frac{m-\ell+1}{2},\frac{\ell+1}{2}\right)\right\}$$

we have

$$\left\|H_{r,n}^{[m]}\right\|_1 \leq \frac{D\xi_n^m}{\left(1-e^{-\pi^2/\xi_n^2}\right)}\sum_{\ell=0}^{m}\omega_r\left(\partial^{m-\ell,\ell}f,\xi_n\right)_1,$$

which finishes the proof. ∎

Furthermore, we obtain the next result.

Theorem 5.5. *Let* $m,r\in\mathbb{N}$ *and* $f\in C_\pi^{(m)}(\mathbb{D})$. *Then*

$$\left\|W_{r,n}^{[m]}(f)-f\right\|_1 \leq \frac{D\xi_n^m}{\left(1-e^{-\pi^2/\xi_n^2}\right)}\sum_{\ell=0}^{m}\omega_r\left(\partial^{m-\ell,\ell}f,\xi_n\right)_1+E\lambda_n\sum_{i=1}^{[m/2]}\xi_n^{2i}$$

holds for some positive constants D,E *depending on* m,r; E *also depends on* f.

Proof. By (5.12) and subadditivity of L_1-norm, we see that

$$\left\|W_{r,n}^{[m]}(f) - f\right\|_1 \leq \left\|H_{r,n}^{[m]}\right\|_1 + \frac{2\lambda_n}{\xi_n^2} \sum_{i=1}^{[m/2]} \frac{\delta_{2i,r}^{[m]} \gamma_{n,2i}}{(2i)!}$$

$$\times \left\{ \sum_{\ell=0}^{2i} \binom{2i}{2i-\ell} \left\|\partial^{2i-\ell,\ell} f\right\|_1 B\left(\frac{2i-\ell+1}{2}, \frac{\ell+1}{2}\right) \right\}$$

$$\leq \left\|H_{r,n}^{[m]}\right\|_1 + \lambda_n \sum_{i=1}^{[m/2]} \frac{\delta_{2i,r}^{[m]} \xi_n^{2i}}{(i+1)\ldots(2i)}$$

$$\times \left\{ \sum_{\ell=0}^{2i} \binom{2i}{2i-\ell} \left\|\partial^{2i-\ell,\ell} f\right\|_1 B\left(\frac{2i-\ell+1}{2}, \frac{\ell+1}{2}\right) \right\}.$$

Now by letting

$$E := \max_{1 \leq i \leq [m/2]} \left\{ \frac{\delta_{2i,r}^{[m]}}{(i+1)\ldots(2i)} \sum_{\ell=0}^{2i} \binom{2i}{2i-\ell} \left\|\partial^{2i-\ell,\ell} f\right\|_1 B\left(\frac{2i-\ell+1}{2}, \frac{\ell+1}{2}\right) \right\},$$

we have

$$\left\|W_{r,n}^{[m]}(f) - f\right\|_1 \leq \left\|H_{r,n}^{[m]}\right\|_1 + E\lambda_n \sum_{i=1}^{[m/2]} \xi_n^{2i},$$

by Theorem 5.4 now claim is proved. ∎

5.2.2 Estimates in the Case of $m = 0$

We now focus on the estimation in the case of $m = 0$. We first obtain the next result.

Theorem 5.6. *Let $r \in \mathbb{N}$ and $p, q > 1$ such that $\frac{1}{p} + \frac{1}{q} = 1$. Then, for every $f \in L_p(\mathbb{D})$ and 2π-periodic per coordinate, the following inequality*

$$\left\|W_{r,n}^{[0]}(f) - f\right\|_p \leq \frac{K\omega_r(f, \xi_n)_p}{\left(1 - e^{-\pi^2/\xi_n^2}\right)^{\frac{1}{p}}},$$

holds for some positive constant K depending on p, r.

Proof. By (5.1), (5.3) and (5.5), we can write

$$W_{r,n}^{[0]}(f;x,y) - f(x,y)$$

$$= \frac{\lambda_n}{\xi_n^2} \iint_{\mathbb{D}} \left\{ \sum_{j=1}^{r} (-1)^{r-j} \binom{r}{j} (f(x+sj, y+tj) - f(x,y)) \right\}$$

$$\times e^{-(s^2+t^2)/\xi_n^2} ds\,dt$$

$$= \frac{\lambda_n}{\xi_n^2} \iint_{\mathbb{D}} \left[\sum_{j=1}^{r} \left((-1)^{r-j} \binom{r}{j} f(x+sj, y+tj) \right) + (-1)^r \binom{r}{0} f(x,y) \right]$$

$$\times e^{-(s^2+t^2)/\xi_n^2} ds\,dt$$

$$= \frac{\lambda_n}{\xi_n^2} \iint_{\mathbb{D}} \left\{ \sum_{j=0}^{r} (-1)^{r-j} \binom{r}{j} f(x+sj, y+tj) \right\} e^{-(s^2+t^2)/\xi_n^2} ds\,dt.$$

Also, by (5.10), we have

$$W_{r,n}^{[0]}(f;x,y) - f(x,y) = \frac{\lambda_n}{\xi_n^2} \iint_{\mathbb{D}} \Delta_{s,t}^r (f(x,y)) e^{-(s^2+t^2)/\xi_n^2} ds\,dt,$$

which gives

$$\left| W_{r,n}^{[0]}(f;x,y) - f(x,y) \right| \leq \frac{\lambda_n}{\xi_n^2} \iint_{\mathbb{D}} \left| \Delta_{s,t}^r (f(x,y)) \right| e^{-(s^2+t^2)/\xi_n^2} ds\,dt. \quad (5.16)$$

Hence, we obtain

$$\iint_{\mathbb{D}} \left| W_{r,n}^{[0]}(f;x,y) - f(x,y) \right|^p dx\,dy$$

$$\leq \frac{\lambda_n^p}{\xi_n^{2p}} \iint_{\mathbb{D}} \left(\iint_{\mathbb{D}} \left| \Delta_{s,t}^r (f(x,y)) \right| e^{-(s^2+t^2)/\xi_n^2} ds\,dt \right)^p dx\,dy.$$

Now, we get from Hölder's inequality for bivariate integrals that

$$\iint_{\mathbb{D}} \left| W_{r,n}^{[0]}(f;x,y) - f(x,y) \right|^p dx\,dy$$

$$\leq \frac{\lambda_n^p}{\xi_n^{2p}} \iint_{\mathbb{D}} \left\{ \left(\iint_{\mathbb{D}} \left| \Delta_{s,t}^r (f(x,y)) \right|^p e^{-(s^2+t^2)/\xi_n^2} ds\,dt \right) \right.$$

$$\left. \times \left(\iint_{\mathbb{D}} e^{-(s^2+t^2)/\xi_n^2} ds\,dt \right)^{p/q} \right\} dx\,dy.$$

Then, using (5.6), we can write

$$\iint\limits_{\mathbb{D}} \left| W_{r,n}^{[0]}(f;x,y) - f(x,y) \right|^p dxdy$$

$$\leq \frac{\lambda_n^p}{\xi_n^{2p}} \left(\pi\xi_n^2 \left(1 - e^{-\pi^2/\xi_n^2}\right) \right)^{\frac{p}{q}} \iint\limits_{\mathbb{D}} \left(\iint\limits_{\mathbb{D}} \left| \Delta_{s,t}^r (f(x,y)) \right|^p e^{-(s^2+t^2)/\xi_n^2} dsdt \right) dxdy$$

$$= \frac{\lambda_n}{\xi_n^2} \iint\limits_{\mathbb{D}} \left(\iint\limits_{\mathbb{D}} \left| \Delta_{s,t}^r (f(x,y)) \right|^p e^{-(s^2+t^2)/\xi_n^2} dsdt \right) dxdy.$$

Thus, by (5.15), we get

$$\iint\limits_{\mathbb{D}} \left| W_{r,n}^{[0]}(f;x,y) - f(x,y) \right|^p dxdy$$

$$\leq \frac{\lambda_n}{\xi_n^2} \iint\limits_{\mathbb{D}} \omega_r \left(f, \sqrt{s^2+t^2} \right)_p^p e^{-(s^2+t^2)/\xi_n^2} dsdt$$

$$= \frac{4\lambda_n}{\xi_n^2} \iint\limits_{\mathbb{D}_1} \omega_r \left(f, \sqrt{s^2+t^2} \right)_p^p e^{-(s^2+t^2)/\xi_n^2} dsdt$$

$$\leq \frac{4\lambda_n}{\xi_n^2} \omega_r \left(f, \xi_n \right)_p^p \iint\limits_{\mathbb{D}_1} \left(1 + \frac{\sqrt{s^2+t^2}}{\xi_n} \right)^{rp} e^{-(s^2+t^2)/\xi_n^2} dsdt,$$

where $\mathbb{D}_1$ is given by (5.14). After some calculations, we derive that

$$\iint\limits_{\mathbb{D}} \left| W_{r,n}^{[0]}(f;x,y) - f(x,y) \right|^p dxdy$$

$$\leq \frac{4\lambda_n}{\xi_n^2} \omega_r \left(f, \xi_n \right)_p^p \int_0^{\pi/2} \int_0^\pi \left(1 + \frac{\rho}{\xi_n} \right)^{rp} e^{-\rho^2/\xi_n^2} \rho d\rho d\theta$$

$$= \frac{2\omega_r \left(f, \xi_n \right)_p^p}{1 - e^{-\pi^2/\xi_n^2}} \int_0^{\pi/\xi_n} (1+u)^{rp} e^{-u^2} u du$$

$$\leq \frac{2\omega_r \left(f, \xi_n \right)_p^p}{1 - e^{-\pi^2/\xi_n^2}} \left(\int_0^\infty (1+u)^{rp+1} e^{-u^2} du \right).$$

Therefore, we get

$$\left\| W_{r,n}^{[0]}(f) - f \right\|_p \leq \frac{K\omega_r \left(f, \xi_n \right)_p}{\left(1 - e^{-\pi^2/\xi_n^2} \right)^{1/p}},$$

where

$$K := K(p,r) = \left(2 \int_0^\infty (1+u)^{rp+1} e^{-u^2} du \right)^{\frac{1}{p}}.$$

The proof is completed. ∎

Finally we get an estimation in the case of $p = 1$ and $m = 0$.

Theorem 5.7. *For every $f \in L_1(\mathbb{D})$ and 2π-periodic per coordinate, we get*

$$\left\| W_{r,n}^{[0]}(f) - f \right\|_1 \leq \frac{L \omega_r(f, \xi_n)_1}{1 - e^{-\pi^2/\xi_n^2}}$$

for some positive constant L depending on r.

Proof. By (5.16), we easily see that

$$\left\| W_{r,n}^{[0]}(f) - f \right\|_1 = \iint_{\mathbb{D}} \left| W_{r,n}^{[0]}(f; x, y) - f(x,y) \right| dxdy$$

$$\leq \frac{\lambda_n}{\xi_n^2} \iint_{\mathbb{D}} \left(\iint_{\mathbb{D}} \left| \Delta_{s,t}^r (f(x,y)) \right| e^{-(s^2+t^2)/\xi_n^2} ds dt \right) dxdy$$

$$= \frac{\lambda_n}{\xi_n^2} \iint_{\mathbb{D}} \left(\iint_{\mathbb{D}} \left| \Delta_{s,t}^r (f(x,y)) \right| dxdy \right) e^{-(s^2+t^2)/\xi_n^2} ds dt$$

$$\leq \frac{\lambda_n}{\xi_n^2} \iint_{\mathbb{D}} \omega_r \left(f, \sqrt{s^2+t^2} \right)_1 e^{-(s^2+t^2)/\xi_n^2} ds dt$$

$$= \frac{4\lambda_n}{\xi_n^2} \iint_{\mathbb{D}_1} \omega_r \left(f, \sqrt{s^2+t^2} \right)_1 e^{-(s^2+t^2)/\xi_n^2} ds dt,$$

where $\mathbb{D}_1$ is given by (5.14). Now using (5.15), we have

$$\left\| W_{r,n}^{[0]}(f) - f \right\|_1 \leq \frac{4\lambda_n \omega_r(f, \xi_n)_1}{\xi_n^2} \iint_{\mathbb{D}_1} \left(1 + \frac{\sqrt{s^2+t^2}}{\xi_n} \right)^r e^{-(s^2+t^2)/\xi_n^2} ds dt$$

$$= \frac{4\lambda_n \omega_r(f, \xi_n)_1}{\xi_n^2} \int_0^{\pi/2} \int_0^\pi \left(1 + \frac{\rho}{\xi_n} \right)^r e^{-\rho^2/\xi_n^2} \rho d\rho d\theta$$

$$= \frac{2\omega_r(f, \xi_n)_1}{1 - e^{-\pi^2/\xi_n^2}} \int_0^{\pi/\xi_n} (1+u)^r e^{-u^2} u du$$

$$\leq \frac{2\omega_r(f, \xi_n)_1}{1 - e^{-\pi^2/\xi_n^2}} \left(\int_0^\infty (1+u)^{r+1} e^{-u^2} du \right).$$

Then, the last inequality implies that

$$\left\| W_{r,n}^{[0]}(f) - f \right\|_1 \leq \frac{L\omega_r(f, \xi_n)_1}{1 - e^{-\pi^2/\xi_n^2}},$$

where

$$L := L(r) = 2 \int_0^\infty (1+u)^{r+1} e^{-u^2} du.$$

The proof is done. ∎

5.3 Statistical L_p-Approximation by the Operators

By the right continuity of $\omega_r(f; \cdot)_p$ at zero we first obtain the following result.

Lemma 5.8. *Let $A = [a_{jn}]$ be a non-negative regular summability matrix, and let (ξ_n) be a sequence of positive real numbers for which*

$$st_A - \lim_n \xi_n = 0 \tag{5.17}$$

holds. Then, for every $f \in C_{2\pi}^{(m)}(\mathbb{D})$, $m \in \mathbb{N}_0$, we get

$$st_A - \lim_n \omega_r(f; \xi_n)_p = 0, \quad 1 \leq p < \infty. \tag{5.18}$$

5.3.1 Statistical L_p-Approximation in the Case of $m \in \mathbb{N}$

The next result is a direct consequence of Theorems 5.3 and 5.5.

Corollary 5.9. *Let $1 \leq p < \infty$ and $m \in \mathbb{N}$. Then, for every $f \in C_{2\pi}^{(m)}(\mathbb{D})$, we get*

$$\left\| W_{r,n}^{[m]}(f) - f \right\|_p \leq \frac{M_1 \xi_n^m}{\left(1 - e^{-\pi^2/\xi_n^2}\right)^{\frac{1}{p}}} \left\{ \sum_{\ell=0}^m \left(\omega_r \left(\partial^{m-\ell,\ell} f, \xi_n \right)_p \right)^p \right\}^{\frac{1}{p}} + M_2 \lambda_n \sum_{i=1}^{[m/2]} \xi_n^{2i}$$

for some positive constants M_1, M_2 depending on m, p, q, r, where

$$M_1 := \begin{cases} D \text{ (as in Theorem 5.5) if } p = 1 \\ C \text{ (as in Theorem 5.3) if } 1 < p < \infty \text{ with } (1/p) + (1/q) = 1 \end{cases}$$

and

$$M_2 := \begin{cases} E \text{ (as in Theorem 5.5) if } p = 1 \\ B \text{ (as in Theorem 5.3) if } 1 < p < \infty \text{ with } (1/p) + (1/q) = 1. \end{cases}$$

Now we can get the first statistical L_p-approximation result.

Theorem 5.10. *Let $m, r \in \mathbb{N}$ and $A = [a_{jn}]$ be a non-negative regular summability matrix, and let (ξ_n) be a sequence of positive real numbers for which (5.17) holds. Then, for any $f \in C_{2\pi}^{(m)}(\mathbb{D})$, we get*

$$st_A - \lim_n \left\| W_{r,n}^{[m]}(f) - f \right\|_p = 0. \tag{5.19}$$

Proof. From (5.17) and Lemma 5.8 we can write

$$st_A - \lim_n \frac{\xi_n^m}{\left(1 - e^{-\pi^2/\xi_n^2}\right)^{\frac{1}{p}}} = 0,$$

$$st_A - \lim_n \left(\omega_r \left(\partial^{m-\ell,\ell} f, \xi_n \right)_p \right)^p = 0 \text{ for each } \ell = 0, 1, ..., m$$

and

$$st_A - \lim_n \xi_n^{2i} = 0 \text{ for each } i = 1, 2, ..., \left[\frac{m}{2} \right].$$

The above results clearly yield that

$$st_A - \lim_n \frac{\xi_n^m}{\left(1 - e^{-\pi^2/\xi_n^2}\right)^{\frac{1}{p}}} \left\{ \sum_{\ell=0}^{m} \left(\omega_r \left(\partial^{m-\ell,\ell} f, \xi_n \right)_p \right)^p \right\}^{\frac{1}{p}} = 0 \tag{5.20}$$

and

$$st_A - \lim_n \lambda_n \sum_{i=1}^{[m/2]} \xi_n^{2i} = 0. \tag{5.21}$$

Now, for a given $\varepsilon > 0$, consider the following sets:

$$S := \left\{ n \in \mathbb{N} : \left\| W_{r,n}^{[m]}(f) - f \right\|_p \geq \varepsilon \right\},$$

$$S_1 := \left\{ n \in \mathbb{N} : \frac{\xi_n^m}{\left(1 - e^{-\pi^2/\xi_n^2}\right)^{\frac{1}{p}}} \left\{ \sum_{\ell=0}^{m} \left(\omega_r \left(\partial^{m-\ell,\ell} f, \xi_n \right)_p \right)^p \right\}^{\frac{1}{p}} \geq \frac{\varepsilon}{2M_1} \right\},$$

$$S_2 := \left\{ n \in \mathbb{N} : \lambda_n \sum_{i=1}^{[m/2]} \xi_n^{2i} \geq \frac{\varepsilon}{2M_2} \right\}.$$

Then, it follows from Corollary 5.9 that

$$S \subseteq S_1 \cup S_2,$$

which gives, for every $j \in \mathbb{N}$, that

$$\sum_{n \in S} a_{jn} \leq \sum_{n \in S_1} a_{jn} + \sum_{n \in S_2} a_{jn}.$$

Now, taking limit as $j \to \infty$ in the both sides of the last inequality and also using (5.20), (5.21), we deduce that

$$\lim_j \sum_{n \in S} a_{jn} = 0,$$

which implies (5.19). Hence, the proof is finished.   ∎

5.3.2   Statistical L_p-Approximation in the Case of $m = 0$

In this subsection, we first combining Theorems 5.6 and 5.7 as follows:

Corollary 5.11. *Let $1 \le p < \infty$ and $r \in \mathbb{N}$. Then, for every $f \in L_p(\mathbb{D})$ and 2π-periodic per coordinate, we get*

$$\left\| W_{r,n}^{[0]}(f) - f \right\|_p \le \frac{N \omega_r (f, \xi_n)_p}{\left(1 - e^{-\pi^2/\xi_n^2}\right)^{\frac{1}{p}}}$$

for some positive constant N depending on p, r, where

$$N := \begin{cases} L \ (\text{as in Theorem 5.7}) & \text{if } p = 1 \\ K \ (\text{as in Theorem 5.6}) & \text{if } 1 < p < \infty \text{ with } (1/p) + (1/q) = 1. \end{cases}$$

Now we obtain the second statistical L_p-approximation result.

Theorem 5.12. *Let $r \in \mathbb{N}$ and $A = [a_{jn}]$ be a non-negative regular summability matrix, and let (ξ_n) be a sequence of positive real numbers for which (5.17) holds. Then, for any $f \in L_p(\mathbb{D})$ and 2π-periodic per coordinate, we get*

$$st_A - \lim_n \left\| W_{r,n}^{[0]}(f) - f \right\|_p = 0. \tag{5.22}$$

Proof. Setting

$$T_1 := \left\{ n \in \mathbb{N} : \left\| W_{r,n}^{[0]}(f) - f \right\|_p \ge \varepsilon \right\}$$

and

$$T_2 := \left\{ n \in \mathbb{N} : \frac{\omega_r (f, \xi_n)_p}{\left(1 - e^{-\pi^2/\xi_n^2}\right)^{\frac{1}{p}}} \ge \frac{\varepsilon}{N} \right\},$$

it follows from Corollary 5.11 that, for every $\varepsilon > 0$,

$$T_1 \subseteq T_2.$$

Hence, for each $j \in \mathbb{N}$, we obtain that

$$\sum_{n \in T_1} a_{jn} \le \sum_{n \in T_2} a_{jn}.$$

Now, taking $j \to \infty$ in the last inequality and considering Lemma 5.8 and also using the fact that

$$st_A - \lim_n \frac{\omega_r (f, \xi_n)_p}{\left(1 - e^{-\pi^2/\xi_n^2}\right)^{\frac{1}{p}}} = 0,$$

we observe that

$$\lim_j \sum_{n \in T_1} a_{jn} = 0,$$

which implies (5.22).   ∎

5.4   Conclusions

In this section, we give some special cases of the approximation results obtained in this chapter.

In particular, we first consider the case of $A = C_1$, the Cesáro matrix of order one. In this case, from Theorems 5.10 and 5.12 we obtain the next result, immediately.

Corollary 5.13. *Let $m \in \mathbb{N}_0$, $r \in \mathbb{N}$, and let (ξ_n) be a sequence of positive real numbers for which*

$$st - \lim_n \xi_n = 0$$

holds. Then, for all $f \in C_{2\pi}^{(m)} (\mathbb{D})$, we get

$$st - \lim_n \left\| W_{r,n}^{[m]}(f) - f \right\|_p = 0.$$

The second result is the case of $A = I$, the identity matrix. Then, the following approximation theorem is a direct consequence of Theorems 5.10 and 5.12.

Corollary 5.14. *Let $m \in \mathbb{N}_0$, $r \in \mathbb{N}$, and let (ξ_n) be a sequence of positive real numbers for which*

$$\lim_n \xi_n = 0$$

holds. Then, for all $f \in C_{2\pi}^{(m)} (\mathbb{D})$, the sequence $\left(W_{r,n}^{[m]}(f) \right)$ is uniformly convergent to f with respect to the L_p-norm.

Finally, define a sequence (ξ_n) as follows:

$$\xi_n := \begin{cases} 1, & \text{if } n = k^2, \ k = 1, 2, ... \\ \frac{1}{1+n}, & \text{otherwise.} \end{cases} \tag{5.23}$$

Then, it is easy to see that $st - \lim_n \xi_n = 0$. So, if we use this sequence (ξ_n) in the definition of the operator $W_{r,n}^{[m]}$, then we obtain from Corollary 5.13 (or, Theorems 5.10 and 5.12) that

$$st - \lim_n \left\| W_{r,n}^{[m]}(f) - f \right\|_p = 0$$

holds for all $f \in C_{2\pi}^{(m)}(\mathbb{D})$, $1 \le p < \infty$. However, since the sequence (ξ_n) given by (5.23) is non-convergent, the classical L_p-approximation to a function f by the operators $W_{r,n}^{[m]}(f)$ is impossible, i.e., Corollary 5.14 fails for these operators. We should notice that Theorems 5.10 and 5.12, and Corollary 5.13 are also valid when $\lim \xi_n = 0$ because every convergent sequence is A-statistically convergent, and so statistically convergent. But, as in the above example, the theorems given in this chapter work although (ξ_n) is non-convergent. Therefore, this non-trivial example clearly shows that the statistical L_p-approximation results in Theorems 5.10 and 5.12, and also in Corollary 5.13 are stronger than Corollary 5.14.

6

A Baskakov-Type Generalization of Statistical Approximation Theory

In this chapter, with the help of the notion of A-statistical convergence, we get some statistical variants of Baskakov's results on the Korovkin-type approximation theorems. This chapter relies on [16].

6.1 Statistical Korovkin-Type Theorems

In this section, by using the A-statistical convergence, we obtain some approximation results by means of a family of positive linear operators. Now consider the following Baskakov-type linear operators

$$L_n(f;x) = \int\limits_a^b f(y)d\varphi_n(x,y), \quad n \in \mathbb{N} \tag{6.1}$$

defined for $f \in C[a,b]$, where $\varphi_n(x,y)$ is, for every n and for every fixed $x \in [a,b]$, a function of bounded variation with respect to the variable to y on the interval $[a,b]$. We should note that if $\varphi_n(x,y)$ is non-decreasing function with respect to the variable y, then the operators (6.1) will be positive. We denote by E_{2k}, $k \geq 1$, the class of operators (6.1) such that,

G.A. Anastassiou and O. Duman: Towards Intelligent Modeling, ISRL 14, pp. 85–96.
springerlink.com

for each fixed $x \in [a, b]$ and for each $n \in \mathbb{N}$, the integrals

$$I_{2k,n}^{(1)}(y) := \int\limits_a^y \int\limits_a^{y_1} \cdots \int\limits_a^{y_{2k-1}} d\varphi_n(x, y_{2k})...dy_2dy_1 \quad \text{for } a \le y \le x,$$

$$I_{2k,n}^{(2)}(y) := \int\limits_y^b \int\limits_{y_1}^b \cdots \int\limits_{y_{2k-1}}^b d\varphi_n(x, y_{2k})...dy_2dy_1 \quad \text{for } x \le y \le b$$

have a constant sign for all $y \in [a, b]$, which may depend on $n \in \mathbb{N}$. We note that these conditions were first considered by Baskakov [41].

Now we start with the following theorem.

Theorem 6.1. *Let $A = [a_{jn}]$ be a non-negative regular summability matrix. If the operators (6.1) belong to the class E_{2k}, $k \ge 1$, and if*

$$st_A - \lim_n \|L_n(e_i) - e_i\| = 0, \quad i = 0, 1, ..., 2k, \tag{6.2}$$

where $e_i(x) = x^i$, $i = 0, 1, ..., 2k$, then, for every function f having a continuous derivative of order $2k$ on the interval $[a, b]$, we get

$$st_A - \lim_n \|L_n(f) - f\| = 0. \tag{6.3}$$

Proof. By similarity it is enough to prove for the case of $k = 1$. Setting $\Psi(y) = y - x$ for each $x \in [a, b]$, we get

$$L_n(\Psi^2; x) = \int\limits_a^b \Psi^2(y)d\varphi_n(x, y)$$

$$= \int\limits_a^b e_2(y)d\varphi_n(x, y) - 2x \int\limits_a^b e_1(y)d\varphi_n(x, y) + \int\limits_a^b e_0(y)d\varphi_n(x, y),$$

which gives

$$\left\| L_n(\Psi^2) \right\| \le \|L_n(e_2) - e_2\| + 2c \|L_n(e_1) - e_1\| + c^2 \|L_n(e_0) - e_0\|, \tag{6.4}$$

where $c = \max\{|a|, |b|\}$. Hence, for every $\varepsilon > 0$, define the following subsets of the natural numbers:

$$D := \left\{ n : \left\| L_n(\Psi^2) \right\| \ge \varepsilon \right\},$$

$$D_1 := \left\{ n : \|L_n(e_2) - e_2\| \ge \frac{\varepsilon}{3} \right\},$$

$$D_2 := \left\{ n : \|L_n(e_1) - e_1\| \ge \frac{\varepsilon}{6c} \right\},$$

$$D_1 := \left\{ n : \|L_n(e_0) - e_0\| \ge \frac{\varepsilon}{3c^2} \right\}.$$

Then, by (6.4), we obtain that

$$D \subseteq D_1 \cup D_2 \cup D_3.$$

This inclusion implies, for every $j \in \mathbb{N}$, that

$$\sum_{n \in D} a_{jn} \leq \sum_{n \in D_1} a_{jn} + \sum_{n \in D_2} a_{jn} + \sum_{n \in D_3} a_{jn}. \tag{6.5}$$

Now taking $j \to \infty$ in (6.5) and using (6.2) we have

$$\lim_j \sum_{n \in D} a_{jn} = 0,$$

which implies

$$st_A - \lim_n \left\| L_n(\Psi^2) \right\| = 0. \tag{6.6}$$

By hypothesis, it is easy to see that

$$st_A - \lim_n \left\| L_n(\Psi) \right\| = 0. \tag{6.7}$$

On the other hand, breaking up the integral

$$L_n(\Psi^2; x) = \int_a^b (y - x)^2 d\varphi_n(x, y)$$

into two integrals over the intervals $[a, x]$ and $[x, b]$ and integrating twice by parts, we derive that

$$L_n(\Psi^2; x) = 2 \left\{ \int_a^x \int_a^y \int_a^{y_1} d\varphi_n(x, y_2) dy_1 dy + \int_x^b \int_y^b \int_{y_1}^b d\varphi_n(x, y_2) dy_1 dy \right\}. \tag{6.8}$$

By the definition of the class E_2, under the signs of the exterior integrals, we get expressions which have a constant sign. Thus, by (6.6) and (6.8), we obtain that

$$st_A - \lim_n \left\{ \sup_{x \in [a,b]} \left(\int_a^x \left| \int_a^y \int_a^{y_1} d\varphi_n(x, y_2) dy_1 \right| dy + \int_x^b \left| \int_y^b \int_{y_1}^b d\varphi_n(x, y_2) dy_1 \right| dy \right) \right\} = 0. \tag{6.9}$$

Furthermore, since the function f has a continuous second derivative on the interval $[a, b]$, it follows from the well-known Taylor's formula that

$$f(y) = f(x) + f'(x)(y - x) + \int_x^y f''(t)(y - t) dt. \tag{6.10}$$

Now using the linearity of the operators L_n we get

$$L_n(f;x) - f(x) = f(x)\left(L_n(e_0;x) - e_0(x)\right) + f'(x)L_n(\Psi;x) + R_n(x), \quad (6.11)$$

where $R_n(x)$ is given by

$$R_n(x) := \int_a^b \int_x^y f''(t)(y-t)dt d\varphi_n(x,y).$$

Breaking up this integral into two integrals over the intervals $[a,x]$ and $[x,b]$ and integrating twice by parts, we see that

$$R_n(x) = \int_a^x \int_a^y \int_a^{y_1} f''(y)d\varphi_n(x,y_2)dy_1 dy + \int_x^b \int_y^b \int_{y_1}^b f''(y)d\varphi_n(x,y_2)dy_1 dy,$$

which yields that

$$\|R_n\| \le M_1 \sup_{x\in[a,b]} \left(\int_a^x \left| \int_a^y \int_a^{y_1} d\varphi_n(x,y_2)dy_1 \right| dy + \int_x^b \left| \int_y^b \int_{y_1}^b d\varphi_n(x,y_2)dy_1 \right| dy \right),$$
$$(6.12)$$

where $M_1 = \|f''\|$. Thus, by (6.9) and (6.12), we obtain

$$st_A - \lim_n \|R_n\| = 0. \quad (6.13)$$

From (6.11), we can write

$$\|L_n(f) - f\| \le M_2 \|L_n(e_0) - e_0\| + M_3 \|L_n(\Psi)\| + \|R_n\|, \quad (6.14)$$

where $M_2 = \|f\|$ and $M_3 = \|f'\|$. Now, for a given $\varepsilon > 0$, consider the following sets:

$$E := \{n : \|L_n(f) - f\| \ge \varepsilon\},$$
$$E_1 := \left\{n : \|L_n(e_0) - e_0\| \ge \frac{\varepsilon}{3M_2}\right\},$$
$$E_2 := \left\{n : \|L_n(\Psi)\| \ge \frac{\varepsilon}{3M_3}\right\},$$
$$E_3 := \left\{n : \|R_n\| \ge \frac{\varepsilon}{3}\right\}.$$

Then, by (6.14), we see that

$$E \subseteq E_1 \cup E_2 \cup E_3.$$

So, we get, for each $j \in \mathbb{N}$,

$$\sum_{n\in E} a_{jn} \le \sum_{n\in E_1} a_{jn} + \sum_{n\in E_2} a_{jn} + \sum_{n\in E_3} a_{jn}.$$

Letting $j \to \infty$ and using (6.2), (6.7), (6.13) we deduce that

$$\lim_j \sum_{n \in E} a_{jn} = 0,$$

which gives (6.3). Therefore, the proof is finished.   ∎

If one replaces the matrix A by the Cesáro matrix, then the next result follows from Theorem 6.1 immediately.

Corollary 6.2. *If the operators (6.1) belong to the class E_{2k}, $k \geq 1$, and if*

$$st - \lim_n \|L_n(e_i) - e_i\| = 0, \quad i = 0, 1, ..., 2k,$$

then, for every function f having a continuous derivative of order $2k$ on the interval $[a, b]$, we get

$$st - \lim_n \|L_n(f) - f\| = 0.$$

Furthermore, considering the identity matrix instead of any non-negative regular summability matrix in Theorem 6.1, we obtain the following result which was first introduced by Baskakov [41].

Corollary 6.3 ([41]). *If the operators (6.1) belong to the class E_{2k}, $k \geq 1$, and if the sequence $(L_n(e_i))$ is uniformly convergent to e_i ($i = 0, 1, ..., 2k$) on the interval $[a, b]$, then, for every function f with a continuous derivative of order $2k$ on the interval $[a, b]$, the sequence $(L_n(f))$ converges uniformly to f on $[a, b]$.*

Remark 6.4. *Let $A = [a_{jn}]$ be a non-negative regular matrix summability satisfying $\lim_j \max_n a_{jn} = 0$. In this case it is known that A-statistical convergence is stronger than ordinary convergence [92]. So we can choose a sequence (u_n) which is A-statistically convergent to zero but non-convergent. Without loss of generality we may assume that (u_n) is non-negative. Otherwise we replace (u_n) by $(|u_n|)$. Now let L_n be the operators given by (6.1) belonging to the class E_{2k} for $k \geq 1$. Assume further that the operators L_n satisfy the conditions of Corollary 6.3. Consider the following operators*

$$T_n(f; x) = (1 + u_n) L_n(f; x) = (1 + u_n) \int_a^b f(y) d\varphi_n(x, y).$$

Then observe that all conditions of Theorem6.1 hold for the operators T_n. So we have

$$st_A - \lim_n \|T_n(f) - f\| = 0.$$

However, since (u_n) is non-convergent, the sequence $(T_n(f))$ is not uniformly convergent to f (in the usual sense). So, this demonstrates that Theorem 6.1 is a non-trivial generalization of its classical case Corollary 6.3.

We also get

Theorem 6.5. *Let* $A = [a_{jn}]$ *be a non-negative regular summability matrix. If, for the operators* (6.1) *belonging to the class* E_{2k}, $k \geq 1$, *the conditions of Theorem 6.1 hold, and if*

$$\delta_A\left(\left\{n : \int_a^b |d\varphi_n(x,y)| \geq M\right\}\right) = 0 \tag{6.15}$$

for some absolute constant $M > 0$, *then, for every function* $f \in C[a,b]$, *we have*

$$st_A - \lim_n \|L_n(f) - f\| = 0.$$

Proof. Since $\{e_0, e_1, e_2, ...\}$ is a fundamental system of $C[a,b]$ (see, for instance, [93]), for a given $f \in C[a,b]$, we can find a polynomial P given by

$$P(x) = a_0 e_0(x) + a_1 e_1(x) + ... + a_{2k} e_{2k}(x)$$

such that for any $\varepsilon > 0$ the inequality

$$\|f - P\| < \varepsilon \tag{6.16}$$

is satisfied. Setting

$$K := \left\{n : \int_a^b |d\varphi_n(x,y)| \geq M\right\},$$

we see from (6.15) that $\delta_A(\mathbb{N}\backslash K) = 1$. By linearity and monotonicity of the operators L_n, we have

$$\|L_n(f) - L_n(P)\| = \|L_n(f - P)\| \leq \|L_n\| \|f - P\|. \tag{6.17}$$

Since

$$\|L_n\| = \int_a^b |d\varphi_n(x,y)|,$$

it follows from (6.16) and (6.17) that, for all $n \in \mathbb{N}\backslash K$,

$$\|L_n(f) - L_n(P)\| \leq M\varepsilon. \tag{6.18}$$

On the other hand, since

$$L_n(P;x) = a_0 L_n(e_0;x) + a_1 L_n(e_1;x) + ... + a_{2k} L_n(e_{2k};x),$$

we obtain, for every $n \in \mathbb{N}$, that

$$\|L_n(P) - P\| \leq C \sum_{i=0}^{2k} \|L_n(e_i) - e_i\|, \tag{6.19}$$

where $C = \max\{|a_1|, |a_2|, ..., |a_{2k}|\}$. Thus, for every $n \in \mathbb{N} \backslash K$, we get from (6.16), (6.18) and (6.19) that

$$
\begin{aligned}
\|L_n(f) - f\| &\leq \|L_n(f) - L_n(P)\| + \|L_n(P) - P\| + \|f - P\| \\
&\leq (M+1)\varepsilon + C \sum_{i=0}^{2k} \|L_n(e_i) - e_i\|
\end{aligned}
\tag{6.20}
$$

Now, for a given $r > 0$, choose $\varepsilon > 0$ such that $0 < (M+1)\varepsilon < r$. Then define the following sets:

$$
H := \{n \in \mathbb{N} \backslash K : \|L_n(f) - f\| \geq r - (M+1)\varepsilon\},
$$
$$
H_i := \left\{n \in \mathbb{N} \backslash K : \|L_n(e_i) - e_i\| \geq \frac{r - (M+1)\varepsilon}{(2k+1)C}\right\}, \quad i = 0, 1, ..., 2k.
$$

From (6.20), we easily check that

$$
H \subseteq \bigcup_{i=0}^{2k} H_i,
$$

which yields, for every $j \in \mathbb{N}$,

$$
\sum_{n \in H} a_{jn} \leq \sum_{i=0}^{2k} \sum_{n \in H_i} a_{jn}.
\tag{6.21}
$$

If we take limit as $j \to \infty$ and also use the hypothesis (6.2), then we see that

$$
\lim_j \sum_{n \in H} a_{jn} = 0.
$$

So we have

$$
st_A - \lim_n \|L_n(f) - f\| = 0
$$

which completes the proof.   ∎

The following two results are obtained from Theorem 6.5 by taking the Cesáro matrix and the identity matrix, respectively.

Corollary 6.6. *If, for the operators (6.1) belonging to the class E_{2k}, $k \geq 1$, the conditions of Corollary 6.2 hold, and if*

$$
\delta\left(\left\{n : \int_a^b |d\varphi_n(x,y)| \geq M\right\}\right) = 0
$$

for some absolute constant $M > 0$, then, for every function $f \in C[a,b]$, we have

$$
st - \lim_n \|L_n(f) - f\| = 0.
$$

Corollary 6.7 ([41]). *If, for the operators (6.1) belong to the class E_{2k}, $k \geq 1$, the conditions of Corollary 6.3 hold, and if the condition*

$$\int_a^b |d\varphi_n(x,y)| \leq M \tag{6.22}$$

holds, where M is a positive absolute constant, then, for every function $f \in C[a,b]$, the sequence $(L_n(f))$ converges uniformly to f on $[a,b]$.

Remark 6.8. *Observe that the boundedness condition in (6.15), the so-called "statistical uniform boundedness", is weaker than the (classical) uniform boundedness in (6.22). So, Theorem 6.5 is more powerful than Corollary 6.7.*

6.2 Statistical Approximation to Derivatives of Functions

In this section we get some statistical approximations to derivatives of functions by means of the operators L_n defined by (6.1). We should remark that the classical versions of the results obtained here were first proved by Baskakov [41].

We first obtain the next result.

Theorem 6.9. *Let $A = [a_{jn}]$ be a non-negative regular summability matrix. If, for the operators L_n given by (6.1) of the class E_{2k}, $k > 1$, the conditions*

$$st_A - \lim_n \left\| L_n(e_i) - e_i^{(2m)} \right\| = 0, \quad i = 0, 1, ..., 2k, \ m < k, \tag{6.23}$$

hold, then, for every function f with a continuous derivative of order $2m$ on the interval $[a,b]$, we get

$$st_A - \lim_n \left\| L_n(f) - f^{(2m)} \right\| = 0.$$

Proof. By similarity, we only prove for $m = 1$. By (6.11), we can write

$$L_n(f;x) = f(x)L_n(e_0;x) + f'(x)L_n(\Psi;x) + L_n^*(f'';x), \tag{6.24}$$

where

$$L_n^*(f'';x) := R_n(x) = \int_a^b f''(y)d\varphi_n^*(x,y) \tag{6.25}$$

with

$$d\varphi_n^*(x,y) := \begin{cases} \left(\int\limits_a^y \int\limits_a^{y_1} d\varphi_n(x,y_2)dy_1 \right) dy, & \text{if } a \leq y \leq x \\ \left(\int\limits_y^b \int\limits_{y_1}^b d\varphi_n(x,y_2)dy_1 \right) dy, & \text{if } x \leq y \leq b. \end{cases} \tag{6.26}$$

Then, we see that the operators $(L_n^*(f''; x))$ belong to the class E_{2k-2}. By (6.23), we have

$$st_A - \lim_n \|L_n^*(e_i) - e_i\| = 0, \quad i = 0, 1, ..., 2k - 2.$$

Because f'' is continuous on $[a, b]$, it follows from Theorem 6.1 that

$$st_A - \lim_n \|L_n^*(f'') - f''\| = 0. \tag{6.27}$$

Now by (6.24) one can obtain that

$$\|L_n(f) - f''\| \le M_1 \|L_n(e_0)\| + M_2 \|L_n(\Psi)\| + \|L_n^*(f'') - f''\|, \tag{6.28}$$

where $M_1 = \|f\|$ and $M_2 = \|f'\|$. The hypothesis (6.23) implies that

$$st_A - \lim_n \|L_n(e_0)\| = 0, \tag{6.29}$$

$$st_A - \lim_n \|L_n(\Psi)\| = 0. \tag{6.30}$$

For a given $\varepsilon > 0$, define the following sets:

$$U := \{n : \|L_n(f) - f''\| \ge \varepsilon\},$$

$$U_1 := \left\{n : \|L_n(e_0)\| \ge \frac{\varepsilon}{3M_1}\right\},$$

$$U_2 := \left\{n : \|L_n(\Psi)\| \ge \frac{\varepsilon}{3M_2}\right\},$$

$$U_3 := \left\{n : \|L_n^*(f'') - f''\| \ge \frac{\varepsilon}{3}\right\},$$

where L_n^* is given by (6.25). Then, by (6.28), it is easy to check that

$$U \subseteq U_1 \cup U_2 \cup U_3.$$

Then one can obtain, for each $j \in \mathbb{N}$, that

$$\sum_{n \in U} a_{jn} \le \sum_{n \in U_1} a_{jn} + \sum_{n \in U_2} a_{jn} + \sum_{n \in U_3} a_{jn}.$$

Taking limit as $j \to \infty$ on the both sides of the above inequality, we get

$$\lim_j \sum_{n \in U} a_{jn} = 0,$$

which completes the proof.   ∎

One can also obtain the following results.

Corollary 6.10. *If the operators* (6.1) *belong to the class* E_{2k}, $k > 1$, *and if*

$$st - \lim_n \left\| L_n(e_i) - e_i^{(2m)} \right\| = 0, \quad i = 0, 1, ..., 2k, \ m < k,$$

then, for every function f *having a continuous derivative of order* $2m$ *on* $[a, b]$, *we get*

$$st - \lim_n \left\| L_n(f) - f^{(2m)} \right\| = 0.$$

Corollary 6.11 (see [41]). *If the operators* (6.1) *belong to the class* E_{2k}, $k > 1$, *and if the sequence* $(L_n(e_i))$ *is uniformly convergent to* $e_i^{(2m)}$ $(i = 0, 1, ..., 2k$ *and* $m < k)$ *on* $[a, b]$, *then, for every function* f *with a continuous derivative of order* $2m$ *on* $[a, b]$, *the sequence* $(L_n(f))$ *converges uniformly to* $f^{(2m)}$ *on* $[a, b]$.

The next theorem can easily be obtained as in Theorem 6.5.

Theorem 6.12. *Let* $A = [a_{jn}]$ *be a non-negative regular summability matrix. If, for the operators* (6.1) *belonging to the class* E_{2k}, $k > 1$, *the conditions of Theorem 6.9 hold, and if*

$$\delta_A \left(\left\{ n : \int_a^b |d\varphi_n^*(x, y)| \geq M \right\} \right) = 0$$

for some absolute constant $M > 0$, *where* $d\varphi_n^*(x, y)$ *is given by* (6.26), *then, for every function* f *with a continuous derivative of order* $2m$, $m < k$, *on the interval* $[a, b]$, *we have*

$$st_A - \lim_n \left\| L_n(f) - f^{(2m)} \right\| = 0.$$

Now we denote by G_{2k+1}, $k \geq 1$, the class of operators (6.1) such that for each fixed $x \in [a, b]$ and for each $n \in \mathbb{N}$, the following integrals

$$J_{2k+1,n}^{(1)}(y) := \int_a^y \int_a^{y_1} ... \int_a^{y_{2k}} d\varphi_n(x, y_{2k+1})...dy_2 dy_1 \quad \text{for } a \leq y \leq x,$$

$$J_{2k+1,n}^{(2)}(y) := \int_y^b \int_{y_1}^b ... \int_{y_{2k}}^b d\varphi_n(x, y_{2k+1})...dy_2 dy_1 \quad \text{for } x \leq y \leq b$$

have well-defined but opposite signs for all $y \in [a, b]$.

Then we obtain the following approximation theorem.

Theorem 6.13. *Let* $A = [a_{jn}]$ *be a non-negative regular summability matrix. If the operators* (6.1) *belong to the class* G_{2k+1}, $k \geq 1$, *and if*

$$st_A - \lim_n \left\| L_n(e_i) - e_i^{(2m+1)} \right\| = 0, \quad i = 0, 1, ..., 2k + 1, \ m < k \quad (6.31)$$

then, for every function f with a continuous derivative of order 2k + 1 on the interval [a, b], we get

$$st_A - \lim_n \left\| L_n(f) - f^{(2m+1)} \right\| = 0. \tag{6.32}$$

Proof. It is enough to prove for $k = 1$ and $m = 0$. Assume that f has a continuous third derivative on $[a, b]$. Then, we can write, for each $x, y \in [a, b]$, that

$$f(y) = f(x) + \int_x^y f'(t)dt.$$

So using the definition of the operators L_n, we have

$$L_n(f; x) = f(x)L_n(e_0; x) + \int_a^b \int_x^y f'(t)dtd\varphi_n(x, y). \tag{6.33}$$

Breaking up the last integral into two integrals over $[a, x]$ and $[x, b]$ and integrating by parts we see that

$$\int_a^b \int_x^y f'(t)dtd\varphi_n(x, y) = -\int_a^x \int_a^y f'(y)d\varphi_n(x, y_1)dy + \int_x^b \int_y^b f'(y)d\varphi_n(x, y_1)dy$$

$$= \int_a^b f'(y)d\varphi_n^{**}(x, y)$$

$$=: L_n^{**}(f'; x),$$

where

$$d\varphi_n^{**}(x, y) := \begin{cases} -\left(\int_a^y d\varphi_n(x, y_1)\right) dy, & \text{if } a \le y \le x \\ \left(\int_y^b d\varphi_n(x, y_2)\right) dy, & \text{if } x \le y \le b. \end{cases} \tag{6.34}$$

Then, we derive that all conditions of Theorem 6.1 are satisfied for the operators $L_n^{**}(f'; x)$. Since f has a continuous third derivative on $[a, b]$, it follows from Theorem 6.1 that

$$st_A - \lim_n \|L_n^{**}(f') - f'\| = 0. \tag{6.35}$$

On the other hand, by (6.31), it is not hard to see that

$$st_A - \lim_n \|L_n(e_0)\| = 0. \tag{6.36}$$

Now by (6.33) we get

$$L_n(f; x) - f'(x) = f(x)L_n(e_0; x) + L_n^{**}(f'; x) - f'(x),$$

which implies that

$$\|L_n(f) - f'\| \le M_1 \|L_n(e_0)\| + \|L_n^{**}(f') - f'\|,$$

where $M_1 = \|f\|$. For every $\varepsilon > 0$, consider the following sets

$$V := \{n : \|L_n(f) - f'\| \ge \varepsilon\},$$

$$V_1 := \left\{n : \|L_n(e_0)\| \ge \frac{\varepsilon}{2M_1}\right\},$$

$$V_2 := \left\{n : \|L_n^{**}(f') - f'\| \ge \frac{\varepsilon}{2}\right\},$$

we immediately get $V \subseteq V_1 \cup V_2$, which yields, for each $j \in \mathbb{N}$,

$$\sum_{n \in V} a_{jn} \le \sum_{n \in V_2} a_{jn} + \sum_{n \in V_2} a_{jn}.$$

Now taking $j \to \infty$ and using (6.35) and (6.36) we get

$$\lim_j \sum_{n \in V} a_{jn} = 0.$$

The last gives

$$st_A - \lim_n \|L_n(f) - f'\| = 0,$$

which finishes the proof for $l = 0$ and $k = 1$.   ∎

By using a similar idea as in Theorems 6.5 and 6.12, we obtain the following result at once.

Theorem 6.14. *Let $A = [a_{jn}]$ be a non-negative regular summability matrix. If, for the operators (6.1) belonging to the class G_{2k+1}, $k \ge 1$, the conditions of Theorem 6.13 hold, and if*

$$\delta_A\left(\left\{n : \int_a^x \left|J_{2k+1,n}^{(1)}(y)\right| dy + \int_x^b \left|J_{2k+1,n}^{(2)}(y)\right| dy \ge M\right\}\right) = 0$$

for some absolute constant $M > 0$, then, for every function f with a continuous derivative of order $2m + 1$, $m < k$, on the interval $[a, b]$, we get

$$st_A - \lim_n \left\|L_n(f) - f^{(2m+1)}\right\| = 0.$$

Finally, we remark that, as in the previous corollaries, one can easily get the statistical and the classical cases of Theorems 6.12, 6.13 and 6.14 by taking the Cesáro matrix and the identity matrix instead of the non-negative regular matrix $A = [a_{jn}]$.

7

Weighted Approximation in Statistical Sense to Derivatives of Functions

In this chapter, we prove some Korovkin-type approximation theorems providing the statistical weighted convergence to derivatives of functions by means of a class of linear operators acting on weighted spaces. We also discuss the contribution of these results to the approximation theory. This chapter relies on [19].

7.1 Statistical Approximation Theorems on Weighted Spaces

Throughout this section, we consider the following weighted spaces introduced by Efendiev [66]. Let k be a non-negative integer. By $C^{(k)}(\mathbb{R})$ we denote the space of all functions having k-th continuous derivatives on $\mathbb{R}$. Now, let $M^{(k)}(\mathbb{R})$ denote the class of linear operators mapping the set of functions f that are convex of order $(k-1)$ on $\mathbb{R}$, i.e., $f^{(k)}(x) \geq 0$ holds for all $x \in \mathbb{R}$, into the set of all positive functions on $\mathbb{R}$. More precisely, for a fixed non-negative integer k and a linear operator L,

$$L \in M^{(k)}(\mathbb{R}) \Leftrightarrow L(f) \geq 0 \text{ for every function } f \text{ satisfying } f^{(k)} \geq 0. \quad (7.1)$$

If $k = 0$, then $M^{(0)}(\mathbb{R})$ stands for the class of all positive linear operators. Suppose that $\rho : \mathbb{R} \to \mathbb{R}^+ = (0, +\infty)$ is a function such that $\rho(0) = 1$; ρ is increasing on $\mathbb{R}^+$ and decreasing on $\mathbb{R}^-$; and $\lim_{x \to \pm\infty} \rho(x) = +\infty$. In this case, we use the following weighted spaces:

G.A. Anastassiou and O. Duman: Towards Intelligent Modeling, ISRL 14, pp. 97–107.
springerlink.com © Springer-Verlag Berlin Heidelberg 2011

$$C_\rho^{(k)}(\mathbb{R}) = \left\{ f \in C^{(k)}(\mathbb{R}) : \text{for some positive } m_f, \ \left| f^{(k)}(x) \right| \leq m_f \rho(x), \ x \in \mathbb{R} \right\},$$

$$\widetilde{C}_\rho^{(k)}(\mathbb{R}) = \left\{ f \in C_\rho^{(k)}(\mathbb{R}) : \text{for some } k_f, \ \lim_{x \to \pm\infty} \frac{f^{(k)}(x)}{\rho(x)} = k_f \right\},$$

$$\widehat{C}_\rho^{(k)}(\mathbb{R}) = \left\{ f \in \widetilde{C}_\rho^{(k)}(\mathbb{R}) : \ \lim_{x \to \pm\infty} \frac{f^{(k)}(x)}{\rho(x)} = 0 \right\},$$

$$B_\rho(\mathbb{R}) = \left\{ g : \mathbb{R} \to \mathbb{R} : \text{for some positive } m_g, \ |g(x)| \leq m_g \rho(x), \ x \in \mathbb{R} \right\}.$$

As usual, the weighted space $B_\rho(\mathbb{R})$ is endowed with the weighted norm

$$\|g\|_\rho := \sup_{x \in \mathbb{R}} \frac{|g(x)|}{\rho(x)} \quad \text{for } g \in B_\rho(\mathbb{R}).$$

If $k = 0$, then we write $M(\mathbb{R})$, $C_\rho(\mathbb{R})$, $\widetilde{C}_\rho(\mathbb{R})$ and $\widehat{C}_\rho(\mathbb{R})$ instead of $M^{(0)}(\mathbb{R})$, $C_\rho^{(0)}(\mathbb{R})$, $\widetilde{C}_\rho^{(0)}(\mathbb{R})$ and $\widehat{C}_\rho^{(0)}(\mathbb{R})$, respectively.

We first recall that the system of functions $f_0, f_1, ..., f_n$ continuous on an interval $[a, b]$ is called a Tschebyshev system of order n, or T-system, if any polynomial

$$P(x) = a_0 f_0(x) + a_1 f_1(x) + ... + a_n f_n(x)$$

has not more than n zeros in this interval with the condition that the numbers $a_0, a_1, ..., a_n$ are not all equal to zero.

Now, following Theorem 3.5 of Duman and Orhan [64] (see also [62, 80]), we get the next statistical approximation result immediately.

Theorem 7.1. *Let $A = [a_{jn}]$ be a non-negative regular summability matrix, and let $\{f_0, f_1, f_2\}$ be T-system on an interval $[a, b]$. Assume that (L_n) is a sequence of positive linear operators from $C[a, b]$ into itself. If*

$$st_A - \lim_n \|L_n(f_i) - f_i\|_{C[a,b]} = 0, \quad i = 0, 1, 2,$$

then, for all $f \in C[a, b]$, we get

$$st_A - \lim_n \|L_n(f) - f\|_{C[a,b]} = 0,$$

where the symbol $\|\cdot\|_{C[a,b]}$ denotes the usual sup-norm on $C[a, b]$.

We first consider the case of $k = 0$.

Theorem 7.2. *Let $A = [a_{jn}]$ be a non-negative regular summability matrix. Assume that the operators $L_n : C_\rho(\mathbb{R}) \to B_\rho(\mathbb{R})$ belong to the class $M(\mathbb{R})$, i.e., they are positive linear operators. Assume further that the following conditions hold:*

(i) $\{f_0, f_1\}$ and $\{f_0, f_1, f_2\}$ are T-systems on $\mathbb{R}$,

(ii) $\lim\limits_{x \to \pm\infty} \dfrac{f_i(x)}{1 + |f_2(x)|} = 0$ for each $i = 0, 1$,

(iii) $\lim\limits_{x \to \pm\infty} \dfrac{f_2(x)}{\rho(x)} = m_{f_2} > 0$,

(iv) $st_A - \lim_n \|L_n(f_i) - f_i\|_\rho = 0$ for each $i = 0, 1, 2$.

Then, for all $f \in \widetilde{C}_\rho(\mathbb{R})$, we get

$$st_A - \lim \|L_n(f) - f\|_\rho = 0.$$

Proof. Let $f \in \widetilde{C}_\rho(\mathbb{R})$ and define a function g on $\mathbb{R}$ as follows

$$g(y) = m_{f_2}\, f(y) - k_f\, f_2(y), \tag{7.2}$$

where m_{f_2} and k_f are certain constants as in the definitions of the weighted spaces. Then, we easily see that $g \in \widehat{C}_\rho(\mathbb{R})$. Now we first claim that

$$st_A - \lim_n \|L_n(g) - g\|_\rho = 0. \tag{7.3}$$

Since $\{f_0, f_1\}$ is T-system on $\mathbb{R}$, we know from Lemma 2 of [66] that, for each $a \in \mathbb{R}$ satisfying $f_i(a) \neq 0$, $i = 0, 1$, there exists a function $\Phi_a(y)$ such that

$$\Phi_a(a) = 0 \text{ and } \Phi_a(y) > 0 \text{ for } y < a,$$

and the function Φ_a has the following form

$$\Phi_a(y) = \gamma_0(a)f_0(y) + \gamma_1(a)f_1(y), \tag{7.4}$$

where $|\gamma_0(a)| = \left| \dfrac{f_1(a)}{f_0(a)} \right|$, and $|\gamma_1(a)| = 1$. Actually, we define

$$\Phi_a(y) = \begin{cases} F(y), & \text{if } F(y) > 0 \text{ for } y < a \\ -F(y), & \text{if } F(y) < 0 \text{ for } y < a, \end{cases}$$

where

$$F(y) = \dfrac{f_1(a)}{f_0(a)} f_0(y) - f_1(y).$$

Clearly here $F(a) = 0$, and F has no other root by $\{f_0, f_1\}$ being a T-system. On the other hand, by (ii) and (iii), we obtain, for each $i = 0, 1$, that

$$\dfrac{f_i(y)}{\rho(y)} = \dfrac{f_i(y)}{1 + |f_2(y)|} \left(\dfrac{1}{\rho(y)} + \dfrac{|f_2(y)|}{\rho(y)} \right) \to 0 \text{ as } y \to \pm\infty. \tag{7.5}$$

Now using the fact that $g \in \widehat{C}_\rho(\mathbb{R})$ and also considering (7.5) and (iii), for every $\varepsilon > 0$, there exists a positive number u_0 such that the conditions

$$|g(y)| < \varepsilon\rho(y), \tag{7.6}$$
$$|f_i(y)| < \varepsilon\rho(y), \quad i = 0, 1, \tag{7.7}$$
$$\rho(y) < s_0 f_2(y), \quad \text{(for a certain positive constant } s_0), \tag{7.8}$$

hold for all y with $|y| > u_0$. By (7.6)-(7.8), we get that

$$|g(y)| < s_0 \varepsilon f_2(y) \quad \text{whenever} \quad |y| > u_0 \tag{7.9}$$

and, for a fixed $a > u_0$ such that $f_i(a) \neq 0, i = 0, 1$,

$$|g(y)| \leq \frac{M}{m_a} \Phi_a(y) \quad \text{whenever} \quad |y| \leq u_0 \tag{7.10}$$

where

$$M := \max_{|y| \leq u_0} |g(y)| \quad \text{and} \quad m_a := \min_{|y| \leq u_0} \Phi_a(y). \tag{7.11}$$

So, combining (7.9) with (7.10), we get

$$|g(y)| < \frac{M}{m_a} \Phi_a(y) + s_0 \varepsilon f_2(y) \quad \text{for all } y \in \mathbb{R}. \tag{7.12}$$

Now, using linearity and monotonicity of the operators L_n, also considering (7.12) and $|\gamma_1(a)| = 1$, we have

$$|L_n(g; x)| \leq L_n\left(|g(y)|; x\right)$$
$$\leq \frac{M}{m_a} L_n\left(\Phi_a(y); x\right) + \varepsilon s_0 L_n(f_2(y); x)$$
$$= \frac{M}{m_a} \left\{\gamma_0(a) L_n(f_0(y); x) + \gamma_1(a) L_n(f_1(y); x)\right\} + s_0 \varepsilon L_n(f_2(y); x)$$
$$\leq \frac{M}{m_a} \left\{|\gamma_0(a)| |L_n(f_0(y); x) - f_0(x)| + |L_n(f_1(y); x) - f_1(x)|\right\}$$
$$+ \frac{M}{m_a} \left\{\gamma_0(a) f_0(x) + \gamma_1(a) f_1(x)\right\} + \varepsilon s_0 |L_n(f_2(y); x) - f_2(x)|$$
$$+ \varepsilon s_0 f_2(x).$$

So we observe

$$\sup_{|x|>u_0} \frac{|L_n(g(y); x)|}{\rho(x)} \leq \frac{M}{m_a} \left\{|\gamma_0(a)| \sup_{|x|>u_0} \frac{|L_n(f_0(y); x) - f_0(x)|}{\rho(x)}\right.$$
$$\left. + \sup_{|x|>u_0} \frac{|L_n(f_1(y); x) - f_1(x)|}{\rho(x)}\right\}$$
$$+ \frac{M}{m_a} \left\{|\gamma_0(a)| \sup_{|x|>u_0} \frac{|f_0(x)|}{\rho(x)} + \sup_{|x|>u_0} \frac{|f_1(x)|}{\rho(x)}\right\}$$
$$+ \varepsilon s_0 \sup_{|x|>u_0} \frac{|L_n(f_2(y); x) - f_2(x)|}{\rho(x)} + \varepsilon s_0 \sup_{|x|>u_0} \frac{|f_2(x)|}{\rho(x)}.$$

But by (7.5) and (iii), we see that

$$A(\varepsilon) := \frac{M}{m_a} \left\{ |\gamma_0(a)| \sup_{|x|>u_0} \frac{|f_0(x)|}{\rho(x)} + \sup_{|x|>u_0} \frac{|f_1(x)|}{\rho(x)} \right\} + \varepsilon s_0 \sup_{|x|>u_0} \frac{|f_2(x)|}{\rho(x)}$$

is finite for every $\varepsilon > 0$. Call now

$$B(\varepsilon) := \max \left\{ \frac{M|\gamma_0(a)|}{m_a}, \frac{M}{m_a}, s_0\varepsilon \right\},$$

which is also finite for every $\varepsilon > 0$. Then we get

$$\sup_{|x|>u_0} \frac{|L_n(g(y);x)|}{\rho(x)} \leq A(\varepsilon) + B(\varepsilon) \sum_{i=0}^{2} \sup_{|x|>u_0} \frac{|L_n(f_i(y);x) - f_i(x)|}{\rho(x)},$$

which gives that

$$\sup_{|x|>u_0} \frac{|L_n(g(y);x)|}{\rho(x)} \leq A(\varepsilon) + B(\varepsilon) \sum_{i=0}^{2} \|L_n(f_i) - f_i\|_\rho. \tag{7.13}$$

On the other hand, since

$$\|L_n(g) - g\|_\rho \leq \sup_{|x|\leq u_0} \frac{|L_n(g(y);x) - g(x)|}{\rho(x)} + \sup_{|x|>u_0} \frac{|L_n(g(y);x)|}{\rho(x)} + \sup_{|x|>u_0} \frac{|g(x)|}{\rho(x)},$$

it follows from (7.6) and (7.13) that

$$\|L_n(g) - g\|_\rho \leq \varepsilon + A(\varepsilon) + B_1 \|L_n(g) - g\|_{C[-u_0,u_0]}$$
$$+ B(\varepsilon) \sum_{i=0}^{2} \|L_n(f_i) - f_i\|_\rho \tag{7.14}$$

holds for every $\varepsilon > 0$ and all $n \in \mathbb{N}$, where $B_1 = \max_{x\in[-u_0,u_0]} \frac{1}{\rho(x)}$. By (iv), we can write that

$$st_A - \lim_n \|L_n(f_i) - f_i\|_{C[-u_0,u_0]} = 0, \quad i = 0, 1, 2. \tag{7.15}$$

Since $\{f_0, f_1, f_2\}$ is T-system and $g \in C[-u_0, u_0]$, we obtain from (7.15) and Theorem 7.1 that

$$st_A - \lim_n \|L_n(g) - g\|_{C[-u_0,u_0]} = 0. \tag{7.16}$$

Now, for a given $r > 0$, choose $\varepsilon > 0$ such that $0 < \varepsilon + A(\varepsilon) < r$. Then, define the following sets:

$$D := \left\{ n \in \mathbb{N} : \|L_n(g) - g\|_\rho \geq r \right\},$$

$$D_1 := \left\{ n \in \mathbb{N} : \|L_n(g) - g\|_{C[-u_0, u_0]} \geq \frac{r - \varepsilon - A(\varepsilon)}{4B_1} \right\},$$

$$D_2 := \left\{ n \in \mathbb{N} : \|L_n(f_0) - f_0\|_\rho \geq \frac{r - \varepsilon - A(\varepsilon)}{4B(\varepsilon)} \right\},$$

$$D_3 := \left\{ n \in \mathbb{N} : \|L_n(f_1) - f_1\|_\rho \geq \frac{r - \varepsilon - A(\varepsilon)}{4B(\varepsilon)} \right\},$$

$$D_4 := \left\{ n \in \mathbb{N} : \|L_n(f_2) - f_2\|_\rho \geq \frac{r - \varepsilon - A(\varepsilon)}{4B(\varepsilon)} \right\}.$$

From (7.14), we easily see that

$$D \subseteq D_1 \cup D_2 \cup D_3 \cup D_4,$$

which implies

$$\sum_{n \in D} a_{jn} \leq \sum_{n \in D_1} a_{jn} + \sum_{n \in D_2} a_{jn} + \sum_{n \in D_3} a_{jn} + \sum_{n \in D_4} a_{jn}. \tag{7.17}$$

Taking $j \to \infty$ in both sides of the inequality (7.17) and also using (iv) and (7.16) we get

$$\lim_j \sum_{n \in D} a_{jn} = 0.$$

Therefore, we prove (7.3). Now, by (7.2), since $f(y) = \dfrac{1}{m_{f_2}} g(y) + \dfrac{k_f}{m_{f_2}} f_2(y)$, we can write, for all $n \in \mathbb{N}$, that

$$
\begin{aligned}
\|L_n(f) - f\|_\rho &= \left\| L_n \left(\frac{1}{m_{f_2}} g + \frac{k_f}{m_{f_2}} f_2 \right) - \left(\frac{1}{m_{f_2}} g + \frac{k_f}{m_{f_2}} f_2 \right) \right\|_\rho \\
&\leq \frac{1}{m_{f_2}} \|L_n(g) - g\|_\rho + \frac{k_f}{m_{f_2}} \|L_n(f_2) - f_2\|_\rho.
\end{aligned}
\tag{7.18}
$$

Now for a given $r' > 0$, consider the sets

$$E := \left\{ n \in \mathbb{N} : \|L_n(f) - f\|_\rho \geq r' \right\},$$

$$E_1 := \left\{ n \in \mathbb{N} : \|L_n(g) - g\|_\rho \geq \frac{m_{f_2} r'}{2} \right\},$$

$$E_2 := \left\{ n \in \mathbb{N} : \|L_n(f_2) - f_2\|_\rho \geq \frac{m_{f_2} r'}{2k_f} \right\}.$$

Then, (7.18) yields that

$$E \subseteq E_1 \cup E_2.$$

So, we obtain, for all $j \in \mathbb{N}$, that

$$\sum_{n \in E} a_{jn} \leq \sum_{n \in E_1} a_{jn} + \sum_{n \in E_2} a_{jn}. \tag{7.19}$$

Letting $j \to \infty$ in the inequality (7.19), and applying (iv) and (7.3), we immediately deduce

$$\lim_j \sum_{n \in E} a_{jn} = 0,$$

which implies

$$st_A - \lim_n \|L_n(f) - f\|_\rho = 0.$$

The proof is completed. ∎

Now, we consider the case of $k \geq 1$.

Theorem 7.3. *Let $A = [a_{jn}]$ be a non-negative regular summability matrix. Assume that the operators $L_n : C_\rho^{(k)}(\mathbb{R}) \to B_\rho(\mathbb{R})$ belong to the class $M^{(k)}(\mathbb{R})$. Let f_0, f_1, f_2 be functions having k-th continuous derivatives on $\mathbb{R}$. Assume further that the following conditions hold:*

(a) $\{f_0^{(k)}, f_1^{(k)}\}$ *and* $\{f_0^{(k)}, f_1^{(k)}, f_2^{(k)}\}$ *are T-systems on $\mathbb{R}$,*

(b) $\displaystyle \lim_{x \to \pm\infty} \frac{f_i^{(k)}(x)}{1 + \left|f_2^{(k)}(x)\right|} = 0$ *for each $i = 0, 1$,*

(c) $\displaystyle \lim_{x \to \pm\infty} \frac{f_2^{(k)}(x)}{\rho(x)} = m_{f_2}^{(k)} > 0$,

(d) $st_A - \lim_n \left\| L_n(f_i) - f_i^{(k)} \right\|_\rho = 0$ *for each $i = 0, 1, 2$.*

Then, for all $f \in \widetilde{C}_\rho^{(k)}(\mathbb{R})$, we get

$$st_A - \lim_n \left\| L_n(f) - f^{(k)} \right\|_\rho = 0.$$

Proof. We say that $f, g \in \widetilde{C}_\rho^{(k)}(\mathbb{R})$ are equivalent provided that $f^{(k)}(x) = g^{(k)}(x)$ for all $x \in \mathbb{R}$. We denote the equivalent classes of $f \in \widetilde{C}_\rho^{(k)}(\mathbb{R})$ by $[f]$. This means that

$$[f] = d^{-k} d^k f,$$

where d^k denotes the k-th derivative operator, and d^{-k} denotes the k-th inverse derivative operator. Thus, by $\left[\widetilde{C}_\rho^{(k)}(\mathbb{R})\right]$ we denote the equivalent weighted spaces of $\widetilde{C}_\rho^{(k)}(\mathbb{R})$. Then, for $f \in \widetilde{C}_\rho^{(k)}(\mathbb{R})$, consider

$$L_n([f]) = L_n\left(d^{-k} d^k f\right) =: L_n^*(\psi), \tag{7.20}$$

where $f^{(k)} = \psi \in \widetilde{C}_\rho(\mathbb{R})$; and L_n^* is an operator such that $L_n^* = L_n d^{-k}$. Then, we can prove that each L_n^* is a positive linear operator from $\widetilde{C}_\rho(\mathbb{R})$ into $B_\rho(\mathbb{R})$. Indeed, if $\psi \geq 0$, i.e., $f^{(k)} \geq 0$, then since each L_n belongs to the class $M^{(k)}(\mathbb{R})$, it follows from (7.1) that $L_n([f]) \geq 0$, i.e., $L_n^*(\psi) \geq 0$ (see also [66]). Now, for every $x \in \mathbb{R}$, considering

$$\psi_i(x) := f_i^{(k)}(x), \quad i = 0, 1, 2,$$

it follows from $(a) - (d)$ that

$\{\psi_0, \psi_1\}$ and $\{\psi_0, \psi_1, \psi_2\}$ are T-systems on $\mathbb{R}$,

$$\lim_{x \to \pm\infty} \frac{\psi_i(x)}{1 + |\psi_2(x)|} = 0 \text{ for each } i = 0, 1,$$

$$\lim_{x \to \pm\infty} \frac{\psi_2(x)}{\rho(x)} = m_{\psi_2} > 0$$

$$st_A - \lim_n \left\| L_n([f_i]) - f_i^{(k)} \right\|_\rho = st_A - \lim_n \| L_n^*(\psi_i) - \psi_i \|_\rho = 0, \quad i = 0, 1, 2.$$

So, all conditions of Theorem 7.2 hold for the functions ψ_0, ψ_1, ψ_2 and the positive linear operators L_n^* given by (7.20). Therefore, we immediately obtain that

$$st_A - \lim_n \| L_n^*(\psi) - \psi \|_\rho = 0,$$

or equivalently,

$$st_A - \lim_n \left\| L_n(f) - f^{(k)} \right\|_\rho = 0,$$

whence the result. ∎

Finally, we get the following result.

Theorem 7.4. *Assume that conditions* (a), (b) *and* (d) *of Theorem 7.3 hold. Let* $\rho_1 : \mathbb{R} \to \mathbb{R}^+ = (0, +\infty)$ *be a function such that* $\rho_1(0) = 1$; ρ_1 *is increasing on* $\mathbb{R}^+$ *and decreasing on* $\mathbb{R}^-$; *and* $\lim_{x \to \pm\infty} \rho_1(x) = +\infty$. *If*

$$\lim_{x \to \pm\infty} \frac{\rho(x)}{\rho_1(x)} = 0, \tag{7.21}$$

and

$$\lim_{x \to \pm\infty} \frac{f_2^{(k)}(x)}{\rho_1(x)} = m_{f_2}^{(k)} > 0 \tag{7.22}$$

then, for all $f \in C_\rho^{(k)}(\mathbb{R})$, *we get*

$$st_A - \lim_n \left\| L_n(f) - f^{(k)} \right\|_{\rho_1} = 0.$$

Proof. Let $f \in C_\rho^{(k)}(\mathbb{R})$. Since $\dfrac{|f^{(k)}(x)|}{\rho(x)} \leq m_f$ for every $x \in \mathbb{R}$, we have

$$\lim_{x \to \pm\infty} \frac{|f^{(k)}(x)|}{\rho_1(x)} \leq \lim_{x \to \pm\infty} \frac{|f^{(k)}(x)|}{\rho(x)} \frac{\rho(x)}{\rho_1(x)} \leq m_f \lim_{x \to \pm\infty} \frac{\rho(x)}{\rho_1(x)}.$$

Then, by (7.21), we easily see that

$$\lim_{x \to \pm\infty} \frac{f^{(k)}(x)}{\rho_1(x)} = 0,$$

which implies

$$f \in \widehat{C}_{\rho_1}^{(k)}(\mathbb{R}) \subset \widetilde{C}_{\rho_1}^{(k)}(\mathbb{R}).$$

Also observe that, by (7.21), condition (d) of Theorem 7.3 holds for the weight function ρ_1. So, the proof follows from Theorem 7.3 and condition (7.22) immediately.   ∎

7.2 Conclusions

If we replace the matrix $A = [a_{jn}]$ in Theorems 7.3 and 7.4 with the identity matrix, then one can immediately obtain the next results in [66], respectively.

Corollary 7.5 (see [66]). *Let f_0, f_1, f_2 be functions having k-th continuous derivatives on $\mathbb{R}$ such that $\{f_0^{(k)}, f_1^{(k)}\}$ and $\{f_0^{(k)}, f_1^{(k)}, f_2^{(k)}\}$ are T-systems on $\mathbb{R}$. Assume that the operators $L_n : C_\rho^{(k)}(\mathbb{R}) \to B_\rho(\mathbb{R})$ belong to the class $M^{(k)}(\mathbb{R})$. Assume further that the following conditions hold:*

(i) $\displaystyle \lim_{t \to \pm\infty} \frac{f_i^{(k)}(x)}{1 + \left| f_2^{(k)}(x) \right|} = 0 \quad (i = 0, 1),$

(ii) $\displaystyle \lim_{t \to \pm\infty} \frac{f_2^{(k)}(x)}{\rho(x)} = m_{f_2}^{(k)} > 0,$

(iii) $\displaystyle \lim_n \left\| L_n(f_i) - f_i^{(k)} \right\|_\rho = 0 \quad (i = 0, 1, 2).$

Then, for all $f \in \widetilde{C}_\rho^{(k)}(\mathbb{R})$, $\lim_n \left\| L_n(f) - f^{(k)} \right\|_\rho = 0$.

Corollary 7.6 (see [66]). *Assume that conditions (i) and (iii) of Corollary 7.5 are satisfied. If (7.21) and (7.22) hold, then, for all $C_\rho^{(k)}(\mathbb{R})$,*

$$\lim_n \left\| L_n(f) - f^{(k)} \right\|_{\rho_1} = 0.$$

Assume now that (L_n) is a sequence of linear operators satisfying all conditions of Corollary 7.5. Let $A = [a_{jn}]$ be a non-negative regular matrix such that $\lim_j \max_n \{a_{jn}\} = 0$. In this case, we know [92] that A-statistical convergence is stronger than the ordinary convergence. So, we can take a sequence (u_n) that is A-statistically null but non-convergent (in the usual sense). Without loss of generality we can assume that (u_n) is a non-negative; otherwise we would replace (u_n) by $(|u_n|)$. Now define

$$T_n(f; x) := (1 + u_n)L_n(f; x). \tag{7.23}$$

By Corollary 7.5, we obtain, for all $f \in \widetilde{C}_\rho^{(k)}(\mathbb{R})$, that

$$\lim_n \left\| L_n(f) - f^{(k)} \right\|_\rho = 0. \tag{7.24}$$

Since $st_A - \lim u_n = 0$, it follows from (7.23) and (7.24) that

$$st_A - \lim_n \left\| T_n(f) - f^{(k)} \right\|_\rho = 0.$$

However, since (u_n) is non-convergent, the sequence $\left\{ \left\| T_n(f) - f^{(k)} \right\|_\rho \right\}$ does not converge to zero. So, Corollary 7.5 does not work for the operators T_n given by (7.23) while Theorem 7.3 still works. It clearly demonstrates that the results obtained in this chapter are non-trivial generalizations of that of Efendiev [66]. Observe that if one takes $A = C_1$, the Cesáro matrix of order one, then Theorem 1 of [80] is an immediate consequence of Theorem 7.3.

Now, in Theorem 7.4, take $k = 0$ and define the test functions

$$f_i(x) = \frac{x^i \rho(x)}{1 + x^2}, \quad i = 0, 1, 2. \tag{7.25}$$

Then, it is easy to see that $\{f_0, f_1\}$ and $\{f_0, f_1, f_2\}$ are T-systems on $\mathbb{R}$. We also derive that the test functions f_i given by (7.25) satisfy the following conditions.

$$\lim_{x \to \pm\infty} \frac{f_0(x)}{1 + |f_2(x)|} = \lim_{x \to \pm\infty} \frac{\rho(x)}{1 + x^2 + x^2 \rho(x)} = 0,$$

$$\lim_{x \to \pm\infty} \frac{f_1(x)}{1 + |f_2(x)|} = \lim_{x \to \pm\infty} \frac{x\rho(x)}{1 + x^2 + x^2 \rho(x)} = 0,$$

$$\lim_{x \to \pm\infty} \frac{f_2(x)}{\rho(x)} = \lim_{x \to \pm\infty} \frac{x^2}{1 + x^2} = 1.$$

Therefore, with these choices, Theorem 3 of [63] is an immediate consequence of Theorem 7.4 for $k = 0$ as follows:

Corollary 7.7 (see [63]). *Let $A = [a_{jn}]$ be a non-negative regular summability matrix, and let (L_n) be a sequence of positive linear operators from $C_\rho(\mathbb{R})$ into $B_\rho(\mathbb{R})$. Assume that the weight functions ρ and ρ_1 satisfy (7.21). If*

$$st_A - \lim_n \| L_n(f_i) - f_i \|_\rho = 0, \quad i = 0, 1, 2,$$

where the functions f_i is given by (7.25), then, for all $f \in C_\rho(\mathbb{R})$, we get

$$st_A - \lim_n \| L_n(f) - f \|_{\rho_1} = 0.$$

Finally, if we replace the matrix $A = [a_{jn}]$ in Corollary 7.7 with the identity matrix, then we get the next classical weighted approximation result for a sequence of positive linear operators (see [77, 79]).

Corollary 7.8. *Let (L_n) be a sequence of positive linear operators from $C_\rho(\mathbb{R})$ into $B_\rho(\mathbb{R})$. Assume that the weight functions ρ and ρ_1 satisfy (7.21). If*

$$\lim_n \|L_n(f_i) - f_i\|_\rho = 0 , \quad i = 0, 1, 2,$$

where the functions f_i is given by (7.25), then, for all $f \in C_\rho(\mathbb{R})$, we get

$$\lim_n \|L_n(f) - f\|_{\rho_1} = 0.$$

8

Statistical Approximation to Periodic Functions by a General Family of Linear Operators

In this chapter, using A-statistical convergence and also considering some matrix summability methods, we introduce an approximation theorem, which is a non-trivial generalization of Baskakov's result [40] regarding the approximation to periodic functions by a general class of linear operators. This chapter relies on [20].

8.1 Basics

Consider the sequence of linear operators

$$L_n(f;x) = \frac{1}{\pi} \int\limits_{-\pi}^{\pi} f(x+t)U_n(t)dt, \quad f \in C_{2\pi} \text{ and } n = 1, 2, ..., \quad (8.1)$$

where

$$U_n(t) = \frac{1}{2} + \sum_{k=1}^{n} \lambda_k^{(n)} \cos kt.$$

As usual, $C_{2\pi}$ denotes the space of all 2π-periodic and continuous functions on the whole real line, endowed with the norm

$$\|f\|_{C_{2\pi}} := \sup_{x \in \mathbb{R}} |f(x)|, \quad f \in C_{2\pi}.$$

If $U_n(t) \geq 0$, $t \in [0, \pi]$, then the operators (8.1) are positive. In this case, Korovkin [93] obtained the following approximation theorem:

G.A. Anastassiou and O. Duman: Towards Intelligent Modeling, ISRL 14, pp. 109–115.
springerlink.com © Springer-Verlag Berlin Heidelberg 2011

Theorem A (see [93]). *If* $\lim_{n\to\infty} \lambda_1^{(n)} = 1$ *and* $U_n(t) \geq 0$ *for all* $t \in [0, \pi]$ *and* $n \in N$, *then, for all* $f \in C_{2\pi}$,

$$\lim_{n\to\infty} L_n(f; x) = f(x) \quad \text{uniformly with respect to all } x \in \mathbb{R}.$$

Notice that Theorem A is valid for the positive linear operators (8.1) we consider. However, Baskakov [40] proves that an analogous result is also valid for a more general class of linear operators that are not necessarily positive. In this chapter, using the concept of statistical convergence we obtain a generalization of both of the results of Korovkin and Baskakov.

8.2   A Statistical Approximation Theorem for Periodic Case

We denote by E the class of operators L_n as in (8.1) such that the integrals

$$\int_t^{\pi/2} \int_{t_1}^{\pi} U_n(t_2) dt_2 dt_1, \quad 0 \leq t < \frac{\pi}{2},$$

$$\int_{\pi/2}^{t} \int_{t_1}^{\pi} U_n(t_2) dt_2 dt_1, \quad \frac{\pi}{2} \leq t \leq \pi,$$

are non-negative. Obviously, the class E contains the class of positive linear operators L_n with $U_n(t) \geq 0$, $t \in [0, \pi]$.

Now we are ready to state the following main result.

Theorem 8.1. *Let* $A = [a_{jn}]$ *be a non-negative regular summability matrix. If the sequence of operators (8.1) belongs to the class E, and if the following conditions*

(a) $st_A - \lim_n \lambda_1^{(n)} = 1$,

(b) $\delta_A \left(\left\{ n : \|L_n\| = \frac{1}{\pi} \int_{-\pi}^{\pi} |U_n(t)| \, dt > M \right\} \right) = 0$

hold for some $M > 0$, then, for all $f \in C_{2\pi}$, we get

$$st_A - \lim_n \|L_n(f) - f\|_{C_{2\pi}} = 0.$$

Proof. Because the functions $\cos t$ and $U_n(t)$ are even, we can write from (8.1) that

$$1 - \lambda_1^{(n)} = \frac{2}{\pi} \int_0^{\pi} (1 - \cos t) U_n(t) dt.$$

Now integrating twice by parts of the above integral we get

$$1 - \lambda_1^{(n)} = \frac{2}{\pi} \int_0^\pi \sin t \left(\int_t^\pi U_n(t_1) dt_1 \right) dt$$

$$= \frac{2}{\pi} \int_0^\pi \cos t \left(\int_t^{\pi/2} \int_{t_1}^\pi U_n(t_2) dt_2 dt_1 \right) dt.$$

By the hypothesis (a), we observe that

$$st_A - \lim_n \left\{ \int_0^\pi \cos t \left(\int_t^{\pi/2} \int_{t_1}^\pi U_n(t_2) dt_2 dt_1 \right) \right\} dt = 0. \qquad (8.2)$$

Since the operators belong to E, the sign of the term inside the brackets is the same as the function $\cos t$ for all $t \in [0, \pi]$. So, it follows from (8.2) that

$$st_A - \lim_n \left\{ \int_0^\pi \left| \cos t \left(\int_t^{\pi/2} \int_{t_1}^\pi U_n(t_2) dt_2 dt_1 \right) \right| dt \right\} = 0. \qquad (8.3)$$

We now claim that

$$st_A - \lim_n \left\{ \int_0^\pi \left| \int_t^{\pi/2} \int_{t_1}^\pi U_n(t_2) dt_2 dt_1 \right| dt \right\} = 0. \qquad (8.4)$$

To prove it, for any $\varepsilon > 0$, we first choose $\delta = \delta(\varepsilon)$ such that $0 < \delta < \sqrt{\dfrac{\varepsilon}{M\pi}}$. Since

$$\int_0^\pi \left| \int_t^{\pi/2} \int_{t_1}^\pi U_n(t_2) dt_2 dt_1 \right| dt \leq \int_{|t-\pi/2| \leq \delta} \left| \int_t^{\pi/2} \int_{t_1}^\pi U_n(t_2) dt_2 dt_1 \right| dt$$

$$+ \int_{|t-\pi/2| > \delta} \left| \int_t^{\pi/2} \int_{t_1}^\pi U_n(t_2) dt_2 dt_1 \right| dt,$$

we obtain

$$\int_0^\pi \left| \int_t^{\pi/2} \int_{t_1}^\pi U_n(t_2) dt_2 dt_1 \right| dt \leq J_{n,1} + J_{n,2}, \qquad (8.5)$$

where

$$J_{n,1} := \int\limits_{|t-\pi/2|\leq\delta} \left| \int\limits_{t}^{\pi/2} \int\limits_{t_1}^{\pi} U_n(t_2)dt_2dt_1 \right| dt$$

and

$$J_{n,2} := \int\limits_{|t-\pi/2|>\delta} \left| \int\limits_{t}^{\pi/2} \int\limits_{t_1}^{\pi} U_n(t_2)dt_2dt_1 \right| dt$$

Putting, for some $M > 0$,

$$K := \{n : \|L_n\| > M\},$$

we obtain from (b) that $\delta_A(\mathbb{N}\backslash K) = 1$. Also, we see that

$$\left| \int\limits_{t_1}^{\pi} U_n(t_2)dt_2 \right| \leq \frac{M\pi}{2}$$

holds for all $n \in \mathbb{N}\backslash K$ and for all $t_1 \in [0, \pi]$. Since $0 < \delta < \sqrt{\dfrac{\varepsilon}{M\pi}}$, we get

$$J_{n,1} < \varepsilon$$

for every $\varepsilon > 0$ and for all $n \in \mathbb{N}\backslash K$. This implies that

$$\lim_{\substack{n\to\infty \\ (n\in\mathbb{N}\backslash K)}} J_{n,1} = 0$$

Since $\delta_A(\mathbb{N}\backslash K) = 1$, it follows from Theorem B that

$$st_A - \lim_n J_{n,1} = 0. \tag{8.6}$$

On the other hand, we have

$$J_{n,2} \leq \left| \int\limits_{|t-\pi/2|>\delta} \frac{\cos t}{\cos(\pi/2 - \delta)} \left(\int\limits_{t}^{\pi/2} \int\limits_{t_1}^{\pi} U_n(t_2)dt_2dt_1 \right) dt \right|,$$

which gives

$$J_{n,2} \leq \frac{1}{\cos(\pi/2 - \delta)} \int\limits_{0}^{\pi} \cos t \left(\int\limits_{t}^{\pi/2} \int\limits_{t_1}^{\pi} U_n(t_2)dt_2dt_1 \right) dt$$

for all $n \in \mathbb{N}$. By (8.3), it is easy to check that

$$st_A - \lim_n J_{n,2} = 0. \tag{8.7}$$

Now, for a given $r > 0$, consider the sets

$$D := \left\{ n : \int_0^{\pi} \left| \int_t^{\pi/2} \int_{t_1}^{\pi} U_n(t_2)dt_2dt_1 \right| dt \geq r \right\},$$

$$D_1 := \left\{ n : J_{n,1} \geq \frac{r}{2} \right\},$$

$$D_2 := \left\{ n : J_{n,2} \geq \frac{r}{2} \right\}.$$

Then, by (8.5), we immediately see that

$$D \subseteq D_1 \cup D_2,$$

and hence

$$\sum_{n \in D} a_{jn} \leq \sum_{n \in D_1} a_{jn} + \sum_{n \in D_2} a_{jn} \tag{8.8}$$

holds for all $j \in \mathbb{N}$. Letting $j \to \infty$ in both sides of (8.8) and also using (8.6), (8.7), we derive that

$$\lim_j \sum_{n \in D} a_{jn} = 0,$$

which proves the claim (8.4). Now let m be an arbitrary non-negative integer. Since

$$\left| 1 - \lambda_m^{(n)} \right| = \left| \frac{2}{\pi} \int_0^{\pi} (1 - \cos mt) U_n(t) dt \right|$$

$$= \left| \frac{2m^2}{\pi} \int_0^{\pi} \cos mt \left(\int_t^{\pi/2} \int_{t_1}^{\pi} U_n(t_2)dt_2dt_1 \right) dt \right|$$

$$\leq \frac{2m^2}{\pi} \int_0^{\pi} \left| \int_t^{\pi/2} \int_{t_1}^{\pi} U_n(t_2)dt_2dt_1 \right| dt,$$

(8.4) yields, for every $m \geq 0$, that

$$st_A - \lim_n \lambda_m^{(n)} = 1.$$

The operators (8.1) can be written as follows:

$$L_n(f;x) = \frac{1}{\pi} \int_{-\pi}^{\pi} f(t) \left\{ \frac{1}{2} + \sum_{k=1}^{n} \cos k(t - x) \right\} dt,$$

see, e.g., [93, p. 68]. Then we have

$$L_n(1; x) = 1$$

and

$$L_n(\cos kt; x) = \lambda_k^{(n)} \cos kx,$$
$$L_n(\sin kt; x) = \lambda_k^{(n)} \sin kx$$

for $k = 1, 2, ...$, and for all $n \in \mathbb{N}$, see, e.g., [93, p. 69]. Thus, we observe that

$$st_A - \lim_n \|L_n(f_m) - f_m\|_{C_{2\pi}} = 0,$$

where the set $\{f_m : m = 0, 1, 2, ...\}$ denotes the class

$$\{1, \cos x, \sin x, \cos 2x, \sin 2x, ...\}.$$

Since $\{f_0, f_1, f_2, ...\}$ is a fundamental system of $C_{2\pi}$ (see, for instance, [93]), for a given $f \in C_{2\pi}$, we can obtain a trigonometric polynomial P given by

$$P(x) = a_0 f_0(x) + a_1 f_1(x) + ... + a_m f_m(x)$$

such that for any $\varepsilon > 0$ the inequality

$$\|f - P\|_{C_{2\pi}} < \varepsilon \tag{8.9}$$

holds. By linearity of the operators L_n, we get

$$\|L_n(f) - L_n(P)\|_{C_{2\pi}} = \|L_n(f - P)\|_{C_{2\pi}} \le \|L_n\| \|f - P\|_{C_{2\pi}}. \tag{8.10}$$

It follows from (8.9), (8.10) and (b) that, for all $n \in \mathbb{N}\backslash K$,

$$\|L_n(f) - L_n(P)\|_{C_{2\pi}} \le M\varepsilon. \tag{8.11}$$

On the other hand, since

$$L_n(P; x) = a_0 L_n(f_0; x) + a_1 L_n(f_1; x) + ... + a_m L_n(f_m; x),$$

we get, for every $n \in \mathbb{N}$, that

$$\|L_n(P) - P\|_{C_{2\pi}} \le C \sum_{i=0}^{m} \|L_n(f_i) - f_i\|_{C_{2\pi}}, \tag{8.12}$$

where $C = \max\{|a_0|, |a_1|, ..., |a_m|\}$. Thus, for every $n \in \mathbb{N}\backslash K$, we obtain from (8.9), (8.11) and (8.12) that

$$\|L_n(f) - f\|_{C_{2\pi}} \le \|L_n(f) - L_n(P)\|_{C_{2\pi}} + \|L_n(P) - P\|_{C_{2\pi}} + \|f - P\|_{C_{2\pi}}$$
$$\le (M + 1)\varepsilon + C \sum_{i=0}^{m} \|L_n(f_i) - f_i\|_{C_{2\pi}}$$

$$\tag{8.13}$$

Now, for a given $r > 0$, choose $\varepsilon > 0$ such that $0 < (M+1)\varepsilon < r$. Then consider the following sets:

$$E := \left\{ n \in \mathbb{N}\backslash K : \|L_n(f) - f\|_{C_{2\pi}} \geq r \right\},$$
$$E_i := \left\{ n \in \mathbb{N}\backslash K : \|L_n(f_i) - f_i\|_{C_{2\pi}} \geq \frac{r - (M+1)\varepsilon}{(m+1)C} \right\}, \quad i = 0, 1, ..., m.$$

From (8.13), we easily see that

$$E \subseteq \bigcup_{i=0}^{m} E_i,$$

which implies, for every $j \in \mathbb{N}$,

$$\sum_{n \in E} a_{jn} \leq \sum_{i=0}^{m} \sum_{n \in E_i} a_{jn}. \tag{8.14}$$

Letting $j \to \infty$ in both sides of (8.14) and using the hypothesis (a) we derive that

$$\lim_j \sum_{n \in E} a_{jn} = 0.$$

So we get

$$st_A - \lim_n \|L_n(f) - f\|_{C_{2\pi}} = 0.$$

The proof is done.   ∎

If we replace the matrix A with the identity matrix, then Theorem 8.1 reduces to Baskakov's result (see [40, Theorem 1]). We also see that if the matrix $A = [a_{jn}]$ satisfies the condition $\lim_j \max_n \{a_{jn}\} = 0$, then Baskakov's result does not necessarily hold while Theorem 8.1 still holds. Furthermore, taking the Cesáro matrix C_1 instead of A, one can get the statistical version of Theorem 8.1.

9

Relaxing the Positivity Condition of Linear Operators in Statistical Korovkin Theory

In this chapter, we relax the positivity condition of linear operators in the Korovkin-type approximation theory via the concept of statistical convergence. Especially, we prove some Korovkin-type approximation theorems providing the statistical convergence to derivatives of functions by means of a class of linear operators. This chapter relies on [18].

9.1 Statistical Korovkin-Type Results

In recent years, by relaxing this positivity condition on linear operators, various approximation theorems have also been obtained. For example, in [51], it was considered linear operators acting from positive and convex functions into positive functions, and from positive and concave functions into concave functions, and also from positive and increasing functions into increasing functions. Some related results may also be found in the papers [2, 43, 91]. However, almost all results in the classical theory are based on the validity of the ordinary limit. In this section, by using the notion of statistical convergence, we obtain various Korovkin-type theorems for a sequence of linear operators under some appropriate conditions rather than the positivity condition although the classical limit fails.

Let k be a non-negative integer. As usual, by $C^k[0, 1]$ we denote the space of all k-times continuously differentiable functions on $[0, 1]$ endowed with the sup-norm $\|\cdot\|$. Then, throughout this section we consider the following

G.A. Anastassiou and O. Duman: Towards Intelligent Modeling, ISRL 14, pp. 117–129.
springerlink.com   © Springer-Verlag Berlin Heidelberg 2011

function spaces:

$$\mathcal{A} := \{f \in C^2[0,1] : f \geq 0\},$$
$$\mathcal{B} := \{f \in C^2[0,1] : f'' \geq 0\},$$
$$\mathcal{C} := \{f \in C^2[0,1] : f'' \leq 0\},$$
$$\mathcal{D} := \{f \in C^1[0,1] : f \geq 0\},$$
$$\mathcal{E} := \{f \in C^1[0,1] : f' \geq 0\},$$
$$\mathcal{F} := \{f \in C^1[0,1] : f' \leq 0\}.$$
$$\mathcal{G} := \{f \in C[0,1] : f \geq 0\}.$$

We also consider the test functions

$$e_i(y) = y^i, \quad i = 0, 1, 2, \ldots$$

Then we present the following results.

Theorem 9.1. *Let $A = [a_{jn}]$ be a non-negative regular summability matrix, and let (L_n) be a sequence of linear operators mapping $C^2[0,1]$ onto itself. Assume that*

$$\delta_A\left(\{n \in \mathbb{N} : L_n\left(\mathcal{A} \cap \mathcal{B}\right) \subset \mathcal{A}\}\right) = 1. \tag{9.1}$$

Then

$$st_A - \lim_n \|L_n(e_i) - e_i\| = 0 \quad for \quad i = 0, 1, 2 \tag{9.2}$$

if and only if

$$st_A - \lim_n \|L_n(f) - f\| = 0 \quad for \ all \ \ f \in C^2[0,1]. \tag{9.3}$$

Proof. The implication (9.3) $\Rightarrow$ (9.2) is clear. Assume that (9.2) holds. Let $x \in [0,1]$ be fixed, and let $f \in C^2[0,1]$. By the boundedness and continuity of f, for every $\varepsilon > 0$, there exists a number $\delta > 0$ such that

$$-\varepsilon - \frac{2M_1\beta}{\delta^2}\varphi_x(y) \leq f(y) - f(x) \leq \varepsilon + \frac{2M_1\beta}{\delta^2}\varphi_x(y) \tag{9.4}$$

holds for all $y \in [0,1]$ and for any $\beta \geq 1$, where $M_1 = \|f\|$ and $\varphi_x(y) = (y - x)^2$. Then, by (9.4), we get that

$$g_\beta(y) := \frac{2M_1\beta}{\delta^2}\varphi_x(y) + \varepsilon + f(y) - f(x) \geq 0$$

and

$$h_\beta(y) := \frac{2M_1\beta}{\delta^2}\varphi_x(y) + \varepsilon - f(y) + f(x) \geq 0$$

hold for all $y \in [0,1]$. So, the functions g_β and h_β belong to $\mathcal{A}$. On the other hand, it is easy to see that, for all $y \in [0,1]$,

$$g_\beta''(y) = \frac{4M_1\beta}{\delta^2} + f''(y)$$

and

$$h_\beta''(y) = \frac{4M_1\beta}{\delta^2} - f''(y).$$

If we choose a number β such that

$$\beta \geq \max\left\{1, \frac{\|f''\|\delta^2}{4M_1}\right\}, \tag{9.5}$$

we obtain that (9.4) holds for such β's and also the functions g_β and h_β belong to $\mathcal{B}$ because of $g_\beta''(y) \geq 0$ and $h_\beta''(y) \geq 0$ for all $y \in [0,1]$. So, we get $g_\beta, h_\beta \in \mathcal{A} \cap \mathcal{B}$ under the condition (9.5). Let

$$K_1 := \{n \in \mathbb{N} : L_n(\mathcal{A} \cap \mathcal{B}) \subset \mathcal{A}\}.$$

By (9.1), it is clear that $\delta_A(K_1) = 1$, and so

$$\delta_A(\mathbb{N}\backslash K_1) = 0. \tag{9.6}$$

Then, we can write

$$L_n(g_\beta; x) \geq 0 \quad \text{and} \quad L_n(h_\beta; x) \geq 0 \quad \text{for every} \quad n \in K_1.$$

Now using the fact that $\varphi_x \in \mathcal{A} \cap \mathcal{B}$ and considering the linearity of L_n, we derive that, for every $n \in K_1$,

$$\frac{2M_1\beta}{\delta^2}L_n(\varphi_x; x) + \varepsilon L_n(e_0; x) + L_n(f; x) - f(x)L_n(e_0; x) \geq 0$$

and

$$\frac{2M_1\beta}{\delta^2}L_n(\varphi_x; x) + \varepsilon L_n(e_0; x) - L_n(f; x) + f(x)L_n(e_0; x) \geq 0,$$

or equivalently

$$-\frac{2M_1\beta}{\delta^2}L_n(\varphi_x; x) - \varepsilon L_n(e_0; x) + f(x)(L_n(e_0; x) - e_0)$$
$$\leq L_n(f; x) - f(x)$$
$$\leq \frac{2M_1\beta}{\delta^2}L_n(\varphi_x; x) + \varepsilon L_n(e_0; x) + f(x)(L_n(e_0; x) - e_0).$$

Then, we obtain

$$|L_n(f; x) - f(x)| \leq \varepsilon + \frac{2M_1\beta}{\delta^2}L_n(\varphi_x; x) + (\varepsilon + |f(x)|)|L_n(e_0; x) - e_0|$$

holds for every $n \in K_1$. The last inequality implies that, for every $\varepsilon > 0$ and $n \in K_1$,

$$\|L_n(f) - f\| \leq \varepsilon + (\varepsilon + M_1)\|L_n(e_0) - e_0\|$$
$$+ \frac{2M_1\beta}{\delta^2}\|L_n(e_2) - e_2\|$$
$$+ \frac{4M_1\beta}{\delta^2}\|L_n(e_1) - e_1\|$$
$$+ \frac{2M_1\beta}{\delta^2}\|L_n(e_0) - e_0\|.$$

Thus, we derive

$$\|L_n(f) - f\| \le \varepsilon + C_1 \sum_{k=0}^{2} \|L_n(e_k) - e_k\| \quad \text{for every } n \in K_1, \qquad (9.7)$$

where $C_1 := \max\left\{\varepsilon + M_1 + \dfrac{2M_1\beta}{\delta^2}, \dfrac{4M_1\beta}{\delta^2}\right\}$. Now, for a given $r > 0$, choose an $\varepsilon > 0$ such that $\varepsilon < r$, and consider the following sets:

$$F := \{n \in \mathbb{N} : \|L_n(f) - f\| \ge r\},$$
$$F_k := \left\{n \in \mathbb{N} : \|L_n(e_k) - e_k\| \ge \frac{r-\varepsilon}{3C_1}\right\}, \quad k = 0, 1, 2.$$

Then, it follows from (9.7) that

$$F \cap K_1 \subset \bigcup_{k=0}^{2} (F_k \cap K_1),$$

which gives, for every $j \in \mathbb{N}$, that

$$\sum_{n \in F \cap K_1} a_{jn} \le \sum_{k=0}^{2} \left(\sum_{n \in F_k \cap K_1} a_{jn}\right) \le \sum_{k=0}^{2} \left(\sum_{n \in F_k} a_{jn}\right) \qquad (9.8)$$

Now, letting $j \to \infty$ in the both-sides of (9.8) and using (9.2), we immediately get that

$$\lim_{j} \sum_{n \in F \cap K_1} a_{jn} = 0. \qquad (9.9)$$

Furthermore, since

$$\sum_{n \in F} a_{jn} = \sum_{n \in F \cap K_1} a_{jn} + \sum_{n \in F \cap (\mathbb{N} \setminus K_1)} a_{jn}$$
$$\le \sum_{n \in F \cap K_1} a_{jn} + \sum_{n \in (\mathbb{N} \setminus K_1)} a_{jn}$$

holds for every $j \in \mathbb{N}$, taking again limit as $j \to \infty$ in the last inequality and using (9.6), (9.9) we conclude that

$$\lim_{j} \sum_{n \in F} a_{jn} = 0,$$

which yields that

$$st_A - \lim_{n} \|L_n(f) - f\| = 0.$$

The proof of theorem is finished. ∎

Theorem 9.2. *Let $A = [a_{jn}]$ be a non-negative regular summability matrix, and let (L_n) be a sequence of linear operators mapping $C^2[0,1]$ onto itself. Assume that*

$$\delta_A\left(\{n \in \mathbb{N} : L_n\left(\mathcal{A} \cap \mathcal{C}\right) \subset \mathcal{C}\}\right) = 1. \tag{9.10}$$

Then

$$st_A - \lim_n \left\|[L_n(e_i)]'' - e_i''\right\| = 0 \quad for \ i = 0,1,2,3,4 \tag{9.11}$$

if and only if

$$st_A - \lim_n \left\|[L_n(f)]'' - f''\right\| = 0 \quad for \ all \ f \in C^2[0,1]. \tag{9.12}$$

Proof. It is enough to prove the implication (9.11) $\Rightarrow$ (9.12). Let $f \in C^2[0,1]$ and $x \in [0,1]$ be fixed. As in the proof of Theorem 9.1, we can write that, for every $\varepsilon > 0$, there exists a $\delta > 0$ such that

$$-\varepsilon + \frac{2M_2\beta}{\delta^2}\sigma_x''(y) \le f''(y) - f''(x) \le \varepsilon - \frac{2M_2\beta}{\delta^2}\sigma_x''(y) \tag{9.13}$$

holds for all $y \in [0,1]$ and for any $\beta \ge 1$, where $\sigma_x(y) = -\dfrac{(y-x)^4}{12} + 1$ and $M_2 = \|f''\|$. Then, consider the following functions on $[0,1]$:

$$u_\beta(y) := \frac{2M_2\beta}{\delta^2}\sigma_x(y) + f(y) - \frac{\varepsilon}{2}y^2 - \frac{f''(x)}{2}y^2,$$

and

$$v_\beta(y) := \frac{2M_2\beta}{\delta^2}\sigma_x(y) - f(y) - \frac{\varepsilon}{2}y^2 + \frac{f''(x)}{2}y^2.$$

By (9.13), we have

$$u_\beta''(y) \le 0 \quad \text{and} \quad v_\beta''(y) \le 0 \quad \text{for all } y \in [0,1],$$

which gives that the functions u_β and v_β belong to $\mathcal{C}$. Observe that $\sigma_x(y) \ge \dfrac{11}{12}$ for all $y \in [0,1]$. Then, the inequality

$$\frac{\left(\pm f(y) + \frac{\varepsilon}{2}y^2 \pm \frac{f''(x)}{2}y^2\right)\delta^2}{2M_2\sigma_x(y)} \le \frac{(M_1 + M_2 + \varepsilon)\delta^2}{M_2}$$

holds for all $y \in [0,1]$, where $M_1 = \|f\|$ and $M_2 = \|f''\|$ as stated before. Now, if we choose a number β such that

$$\beta \ge \max\left\{1, \frac{(M_1 + M_2 + \varepsilon)\delta^2}{M_2}\right\}, \tag{9.14}$$

then inequality (9.13) holds for such β's and

$$u_\beta(y) \ge 0 \quad \text{and} \quad v_\beta(y) \ge 0 \quad \text{for all } y \in [0,1].$$

Hence, we also obtain u_β, $v_\beta \in \mathcal{A}$, which implies that the functions u_β and v_β belong to $\mathcal{A} \cap \mathcal{C}$ under the condition (9.14). Now letting

$$K_2 := \{n \in \mathbb{N} : L_n\left(\mathcal{A} \cap \mathcal{C}\right) \subset \mathcal{C}\},$$

and also using (9.10), we obtain that

$$\delta_A(\mathbb{N} \backslash K_2) = 0. \tag{9.15}$$

Also we have, for every $n \in K_2$,

$$[L_n(u_\beta)]'' \leq 0 \quad \text{and} \quad [L_n(v_\beta)]'' \leq 0.$$

Then, we see, for every $n \in K_2$, that

$$\frac{2M_2\beta}{\delta^2}[L_n(\sigma_x)]'' + [L_n(f)]'' - \frac{\varepsilon}{2}[L_n(e_2)]'' - \frac{f''(x)}{2}[L_n(e_2)]'' \leq 0$$

and

$$\frac{2M_2\beta}{\delta^2}[L_n(\sigma_x)]'' - [L_n(f)]'' - \frac{\varepsilon}{2}[L_n(e_2)]'' + \frac{f''(x)}{2}[L_n(e_2)]'' \leq 0.$$

These inequalities imply that

$$\frac{2M_2\beta}{\delta^2}[L_n(\sigma_x)]''(x) - \frac{\varepsilon}{2}[L_n(e_2)]''(x) + \frac{f''(x)}{2}[L_n(e_2)]''(x) - f''(x)$$
$$\leq [L_n(f)]''(x) - f''(x)$$
$$\leq -\frac{2M_2\beta}{\delta^2}[L_n(\sigma_x)]''(x) + \frac{\varepsilon}{2}[L_n(e_2)]''(x) + \frac{f''(x)}{2}[L_n(e_2)]''(x) - f''(x).$$

Observe now that $[L_n(\sigma_x)]'' \leq 0$ on $[0,1]$ for every $n \in K_2$ because of $\sigma_x \in \mathcal{A} \cap \mathcal{C}$. Using this, the last inequality yields, for every $n \in K_2$, that

$$|[L_n(f)]''(x) - f''(x)| \leq -\frac{2M_2\beta}{\delta^2}[L_n(\sigma_x)]''(x) + \frac{\varepsilon}{2}\left|[L_n(e_2)]''(x)\right|$$
$$+ \frac{|f''(x)|}{2}\left|[L_n(e_2)]''(x) - 2\right|,$$

and hence

$$|[L_n(f)]''(x) - f''(x)| \leq \varepsilon + \frac{\varepsilon + |f''(x)|}{2}\left|[L_n(e_2)]''(x) - e_2''(x)\right|$$
$$+ \frac{2M_2\beta}{\delta^2}[L_n(-\sigma_x)]''(x). \tag{9.16}$$

Now we compute the quantity $[L_n(-\sigma_x)]''$ in inequality (9.16). To see this observe that

$$[L_n(-\sigma_x)]''(x) = \left[L_n\left(\frac{(y-x)^4}{12} - 1\right)\right]''(x)$$

$$= \frac{1}{12}[L_n(e_4)]''(x) - \frac{x}{3}[L_n(e_3)]''(x) + \frac{x^2}{2}[L_n(e_2)]''(x)$$

$$- \frac{x^3}{3}[L_n(e_1)]''(x) + \left(\frac{x^4}{12} - 1\right)[L_n(e_0)]''(x)$$

$$= \frac{1}{12}\{[L_n(e_4)]''(x) - e_4''(x)\} - \frac{x}{3}\{[L_n(e_3)]''(x) - e_3''(x)\}$$

$$+ \frac{x^2}{2}\{[L_n(e_2)]''(x) - e_2''(x)\} - \frac{x^3}{3}\{[L_n(e_1)]''(x) - e_1''(x)\}$$

$$+ \left(\frac{x^4}{12} - 1\right)\{[L_n(e_0)]''(x) - e_0''(x)\}.$$

Combining this with (9.16), for every $\varepsilon > 0$ and $n \in K_2$, we get

$$|[L_n(f)]''(x) - f''(x)| \le \varepsilon + \left(\frac{\varepsilon + |f''(x)|}{2} + \frac{M_2\beta x^2}{\delta^2}\right)|[L_n(e_2)]''(x) - e_2''(x)|$$

$$+ \frac{M_2\beta}{6\delta^2}|[L_n(e_4)]''(x) - e_4''(x)|$$

$$+ \frac{2M_2\beta x}{3\delta^2}|[L_n(e_3)]''(x) - e_3''(x)|$$

$$+ \frac{2M_2\beta x^3}{3\delta^2}|[L_n(e_1)]''(x) - e_1''(x)|$$

$$+ \frac{2M_2\beta}{3\delta^2}\left(1 - \frac{x^4}{12}\right)|[L_n(e_0)]''(x) - e_0''(x)|.$$

Therefore, we derive, for every $\varepsilon > 0$ and $n \in K_2$, that

$$\|[L_n(f)]'' - f''\| \le \varepsilon + C_2 \sum_{k=0}^{4} \|[L_n(e_k)]'' - e_k''\|, \qquad (9.17)$$

where $C_2 := \dfrac{\varepsilon + M_2}{2} + \dfrac{M_2\beta}{\delta^2}$ and $M_2 = \|f''\|$ as stated before. Now, for a given $r > 0$, choose an ε such that $0 < \varepsilon < r$, and define the following sets:

$$G := \{n \in \mathbb{N} : \|[L_n(f)]'' - f''\| \ge r\},$$

$$G_k := \left\{n \in \mathbb{N} : \|[L_n(e_k)]'' - e_k''\| \ge \frac{r-\varepsilon}{5C_2}\right\}, \quad k = 0, 1, 2, 3, 4.$$

In this case, by (9.17),

$$G \cap K_2 \subset \bigcup_{k=0}^{4}(G_k \cap K_2),$$

which gives, for every $j \in \mathbb{N}$, that

$$\sum_{n \in G \cap K_2} a_{jn} \leq \sum_{k=0}^{4} \left(\sum_{n \in G_k \cap K_2} a_{jn} \right) \leq \sum_{k=0}^{4} \left(\sum_{n \in G_k} a_{jn} \right) \qquad (9.18)$$

Taking limit as $j \to \infty$ in the both-sides of (9.18) and using (9.11), we immediately check that

$$\lim_{j} \sum_{n \in G \cap K_2} a_{jn} = 0. \qquad (9.19)$$

Furthermore, if we consider the inequality

$$\sum_{n \in G} a_{jn} = \sum_{n \in G \cap K_2} a_{jn} + \sum_{n \in G \cap (\mathbb{N} \setminus K_2)} a_{jn}$$

$$\leq \sum_{n \in G \cap K_2} a_{jn} + \sum_{n \in (\mathbb{N} \setminus K_2)} a_{jn}$$

and if we take limit as $j \to \infty$, then it follows from (9.15) and (9.19) that

$$\lim_{j} \sum_{n \in G} a_{jn} = 0.$$

Thus, we obtain

$$st_A - \lim_{n} \left\| [L_n(f)]'' - f'' \right\| = 0.$$

The proof is completed. ∎

Theorem 9.3. *Let $A = [a_{jn}]$ be a non-negative regular summability matrix, and let (L_n) be a sequence of linear operators mapping $C^1[0,1]$ onto itself. Assume that*

$$\delta_A \left(\{ n \in \mathbb{N} : L_n (\mathcal{D} \cap \mathcal{E}) \subset \mathcal{E} \} \right) = 1. \qquad (9.20)$$

Then

$$st_A - \lim_{n} \left\| [L_n(e_i)]' - e_i' \right\| = 0 \quad \text{for } i = 0, 1, 2, 3 \qquad (9.21)$$

if and only if

$$st_A - \lim_{n} \left\| [L_n(f)]' - f' \right\| = 0 \quad \text{for all } f \in C^1[0,1]. \qquad (9.22)$$

Proof. It is enough to prove the implication (9.21) $\Rightarrow$ (9.22). Let $f \in C^1[0,1]$ and $x \in [0,1]$ be fixed. Then, for every $\varepsilon > 0$, there exists a positive number δ such that

$$-\varepsilon - \frac{2M_3 \beta}{\delta^2} w_x'(y) \leq f'(y) - f'(x) \leq \varepsilon + \frac{2M_3 \beta}{\delta^2} w_x'(y) \qquad (9.23)$$

holds for all $y \in [0,1]$ and for any $\beta \geq 1$, where $w_x(y) := \dfrac{(y-x)^3}{3} + 1$ and $M_3 := \|f'\|$. Now using the functions defined by

$$\theta_\beta(y) := \frac{2M_3\beta}{\delta^2} w_x(y) - f(y) + \varepsilon y + yf'(x)$$

and

$$\lambda_\beta(y) := \frac{2M_3\beta}{\delta^2} w_x(y) + f(y) + \varepsilon y - yf'(x),$$

we can easily see that θ_β and λ_β belong to $\mathcal{E}$ for any $\beta \geq 1$, i.e. $\theta'_\beta(y) \geq 0$, $\lambda'_\beta(y) \geq 0$. Also, observe that $w_x(y) \geq \dfrac{2}{3}$ for all $y \in [0,1]$. Then, the inequality

$$\frac{(\pm f(y) - \varepsilon y \pm f'(x)y)\delta^2}{2M_3 w_x(y)} \leq \frac{(M_1 + M_3 + \varepsilon)\delta^2}{M_3}$$

holds for all $y \in [0,1]$, where $M_1 = \|f\|$. Now, if we choose a number β such that

$$\beta \geq \max\left\{1, \frac{(M_1 + M_3 + \varepsilon)\delta^2}{M_3}\right\}, \tag{9.24}$$

then inequality (9.23) holds for such β's and

$$\theta_\beta(y) \geq 0 \quad \text{and} \quad \lambda_\beta(y) \geq 0 \quad \text{for all } y \in [0,1],$$

which gives that $\theta_\beta, \lambda_\beta \in \mathcal{D}$. Thus, we have $\theta_\beta, \lambda_\beta \in \mathcal{D} \cap \mathcal{E}$ for any β satisfying (9.24). Let

$$K_3 := \{n \in \mathbb{N} : L_n(\mathcal{D} \cap \mathcal{E}) \subset \mathcal{E}\}.$$

Then, by (9.20), we get

$$\delta_A(\mathbb{N} \backslash K_3) = 0. \tag{9.25}$$

Also we obtain, for every $n \in K_3$,

$$[L_n(\theta_\beta)]' \geq 0 \quad \text{and} \quad [L_n(\lambda_\beta)]' \geq 0.$$

Hence, we derive, for every $n \in K_3$, that

$$\frac{2M_3\beta}{\delta^2}[L_n(w_x)]' - [L_n(f)]' + \varepsilon[L_n(e_1)]' + f'(x)[L_n(e_1)]' \geq 0$$

and

$$\frac{2M_3\beta}{\delta^2}[L_n(w_x)]' + [L_n(f)]' + \varepsilon[L_n(e_1)]' - f'(x)[L_n(e_1)]' \geq 0.$$

Then, we observe, for any $n \in K_3$, that

$$
\begin{aligned}
-\frac{2M_3\beta}{\delta^2}&[L_n(w_x)]'(x) - \varepsilon[L_n(e_1)]'(x) + f'(x)[L_n(e_1)]'(x) - f'(x) \\
&\leq [L_n(f)]'(x) - f'(x) \\
&\leq \frac{2M_3\beta}{\delta^2}[L_n(w_x)]'(x) + \varepsilon[L_n(e_1)]'(x) + f'(x)[L_n(e_1)]'(x) - f'(x),
\end{aligned}
$$

and hence

$$
\begin{aligned}
|[L_n(f)]'(x) - f'(x)| \leq &\varepsilon + (\varepsilon + |f'(x)|) \, |[L_n(e_1)]'(x) - e_1'(x)| \\
&+ \frac{2M_3\beta}{\delta^2}[L_n(w_x)]'(x)
\end{aligned}
\tag{9.26}
$$

holds for every $n \in K_3$ because of the fact that the function w_x belongs to $\mathcal{D} \cap \mathcal{E}$. Since

$$
\begin{aligned}
[L_n(w_x)]'(x) &= \left[L_n\left(\frac{(y-x)^3}{3} + 1 \right) \right]'(x) \\
&= \frac{1}{3}[L_n(e_3)]'(x) - x[L_n(e_2)]'(x) \\
&\quad + x^2[L_n(e_1)]'(x) + \left(1 - \frac{x^3}{3} \right)[L_n(e_0)]'(x) \\
&= \frac{1}{3}\{[L_n(e_3)]'(x) - e_3'(x)\} - x\{[L_n(e_2)]'(x) - e_2'(x)\} \\
&\quad + x^2\{[L_n(e_1)]'(x) - e_1'(x)\} + \left(1 - \frac{x^3}{3} \right)\{[L_n(e_0)]'(x) - e_0'(x)\},
\end{aligned}
$$

it follows from (9.26) that

$$
\begin{aligned}
|[L_n(f)]'(x) - f'(x)| \leq &\varepsilon + \left(\varepsilon + |f'(x)| + \frac{2M_3\beta x^2}{\delta^2} \right) |[L_n(e_1)]'(x) - e_1'(x)| \\
&+ \frac{2M_3\beta}{3\delta^2} |[L_n(e_3)]'(x) - e_3'(x)| \\
&+ \frac{2M_3\beta x}{\delta^2} |[L_n(e_2)]'(x) - e_2'(x)| \\
&+ \frac{2M_3\beta}{\delta^2}\left(1 - \frac{x^3}{3} \right) |[L_n(e_0)]'(x) - e_0'(x)|.
\end{aligned}
$$

Thus, we conclude from the last inequality that

$$
\|[L_n(f)]' - f'\| \leq \varepsilon + C_3 \sum_{k=0}^{3} \|[L_n(e_k)]' - e_k'\|
\tag{9.27}
$$

holds for any $n \in K_3$, where $C_3 := \varepsilon + M_3 + \dfrac{2M_3\beta}{\delta^2}$. Now, for a given $r > 0$, choose an ε such that $0 < \varepsilon < r$, and define the following sets:

$$H := \left\{ n \in \mathbb{N} : \left\| [L_n(f)]' - f' \right\| \geq r \right\},$$

$$H_k := \left\{ n \in \mathbb{N} : \left\| [L_n(e_k)]' - e_k' \right\| \geq \frac{r - \varepsilon}{4C_3} \right\}, \quad k = 0, 1, 2, 3.$$

In this case, by (9.27),

$$H \cap K_3 \subset \bigcup_{k=0}^{3} (H_k \cap K_3),$$

which implies, for every $j \in \mathbb{N}$, that

$$\sum_{n \in H \cap K_3} a_{jn} \leq \sum_{k=0}^{3} \left(\sum_{n \in H_k \cap K_3} a_{jn} \right) \leq \sum_{k=0}^{3} \left(\sum_{n \in H_k} a_{jn} \right) \tag{9.28}$$

Taking $j \to \infty$ in the both-sides of (9.28) and also using (9.21), we observe that

$$\lim_{j} \sum_{n \in H \cap K_3} a_{jn} = 0. \tag{9.29}$$

Now, using the fact that

$$\sum_{n \in H} a_{jn} = \sum_{n \in H \cap K_3} a_{jn} + \sum_{n \in H \cap (\mathbb{N} \setminus K_3)} a_{jn}$$

$$\leq \sum_{n \in H \cap K_3} a_{jn} + \sum_{n \in (\mathbb{N} \setminus K_3)} a_{jn}$$

and taking limit as $j \to \infty$, then it follows from (9.25) and (9.29) that

$$\lim_{j} \sum_{n \in H} a_{jn} = 0.$$

Thus, we obtain that

$$st_A - \lim_{n} \left\| [L_n(f)]' - f' \right\| = 0.$$

Thus, the theorem is proved.   ∎

Theorem 9.4. *Let $A = [a_{jn}]$ be a non-negative regular summability matrix, and let (L_n) be a sequence of linear operators mapping $C[0,1]$ onto itself. Assume that*

$$\delta_A \left(\{ n \in \mathbb{N} : L_n(\mathcal{G}) \subset \mathcal{G} \} \right) = 1. \tag{9.30}$$

Then

$$st_A - \lim_{n} \| L_n(e_i) - e_i \| = 0 \quad for \ i = 0, 1, 2 \tag{9.31}$$

if and only if

$$st_A - \lim_n \|L_n(f) - f\| = 0 \quad \text{for all} \quad f \in C[0,1]. \tag{9.32}$$

Proof. See the remarks in the next section. ∎

9.2 Conclusions

In this section we summarize the results obtained in this chapter and give some applications in order to show the importance of using the statistical approximation in this study.

- In Theorem 9.4, if we consider the condition

$$\{n \in \mathbb{N} : L_n(\mathcal{G}) \subset \mathcal{G}\} = \mathbb{N} \tag{9.33}$$

  instead of (9.30), then we see that the linear operators L_n are positive for each $n \in \mathbb{N}$. In this case, Theorem 9.4 is an A-statistical version of Theorem 1 of [80], and the proof follows immediately. Actually, as in the previous proofs, we can show that

$$(9.31) \Leftrightarrow (9.32)$$

  although the weaker condition (9.30) holds. Because of similarity, we omit the proof of Theorem 9.4. Here, condition (9.30) gives that L_n does not need to be positive for each $n \in \mathbb{N}$, but it is enough to be positive for each $n \in K$ with $\delta_A(K) = 1$. Observe that condition (9.30), which is weaker than (9.33), can be applied to many well-known results regarding statistical approximation of positive linear operators, such as Theorem 3 of [63], Theorems 2.1 and 2.2 of [68], Theorem 2.1 of [67] and Theorem 1 of [56].
- We can easily observe that all of the theorems in this chapter are also valid for any compact subset of $\mathbb{R}$ instead of the unit interval $[0,1]$.
- In Theorems 9.1-9.3, if we replace the matrix A by the identity matrix and also if we consider the conditions

$$\{n \in \mathbb{N} : L_n(\mathcal{A} \cap \mathcal{B}) \subset \mathcal{A}\} = \mathbb{N}, \tag{9.34}$$
$$\{n \in \mathbb{N} : L_n(\mathcal{A} \cap \mathcal{C}) \subset \mathcal{C}\} = \mathbb{N}, \tag{9.35}$$
$$\{n \in \mathbb{N} : L_n(\mathcal{D} \cap \mathcal{E}) \subset \mathcal{E}\} = \mathbb{N} \tag{9.36}$$

instead of the conditions (9.1), (9.10) and (9.20), respectively, then we obtain Propositions 1-3 of [51]. Indeed, for example, suppose that A is the identity matrix and (9.34) holds. In this case, since A-statistical convergence coincides with the ordinary convergence, the conditions (9.2) and (9.3) hold with respect to the classical limit operator. Also, according to (9.34), for each $n \in \mathbb{N}$, the linear operators L_n in Theorem 9.1 map positive and convex functions onto positive functions. Hence, we have Proposition 1 of [51].

- Theorem 9.3 is valid if we replace the condition (9.20) by

$$\delta_A (\{n \in \mathbb{N} : L_n (\mathcal{D} \cap \mathcal{F}) \subset \mathcal{F}\}) = 1.$$

To prove this, it is enough to consider the function $\psi_x(y) = -\frac{(y-x)^3}{3} + 1$ instead of $w_x(y)$ defined in the proof of Theorem 9.3.

- The next example clearly shows that the statistical approximation results obtained here are more applicable than the classical ones. Take $A = C_1$ and define the linear operators L_n on $C^2[0,1]$ as follows:

$$L_n(f;x) = \begin{cases} -x^2 & \text{if } n = m^2 \ (m \in \mathbb{N}) \\ B_n(f;x); & \text{otherwise,} \end{cases} \qquad (9.37)$$

where the operators $B_n(f;x)$ denote the Bernstein polynomials. Then, we see that

$$\begin{aligned} \delta_{C_1} (\{n \in \mathbb{N} : L_n (\mathcal{A} \cap \mathcal{B}) \subset \mathcal{A}\}) &= \delta (\{n \in \mathbb{N} : L_n (\mathcal{A} \cap \mathcal{B}) \subset \mathcal{A}\}) \\ &= \delta (\{n \neq m^2 : m \in \mathbb{N}\}) \\ &= 1. \end{aligned}$$

Also we get, for each $i = 0, 1, 2$,

$$st_{C_1} - \lim_n \|L_n(e_i) - e_i\| = st - \lim_n \|L_n(e_i) - e_i\| = 0.$$

Then, it follows from Theorem 9.1 that, for all $f \in C^2[0,1]$,

$$st_{C_1} - \lim_n \|L_n(f) - f\| = 0.$$

However, for the function $e_0 = 1$, since

$$L_n(e_0;x) := \begin{cases} -x^2 & \text{if } n = m^2 \ (m \in \mathbb{N}) \\ 1 & \text{otherwise,} \end{cases}$$

we obtain, for all $x \in [0,1]$, that the sequence $(L_n(e_0;x))$ is non-convergent. This shows that Proposition 1 of [51] does not work while Theorem 9.1 still works for the operators L_n defined by (9.37).

10

Statistical Approximation Theory for Stochastic Processes

In this chapter, we present strong Korovkin-type approximation theorems for stochastic processes via the concept of A-statistical convergence. This chapter relies on [31].

10.1 Statistical Korovkin-Type Results for Stochastic Processes

Let $m \in \mathbb{N}_0$, the set of all non-negative integers. As usual, by $C^m[a, b]$, $a < b$, we denote the space of all k-times continuously differentiable functions on $[a, b]$ endowed with the usual sup-norm $\|\cdot\|$. Then, throughout this section we consider the following concepts and assumptions (*cf.* [3, 11]):

(a) Let (L_n) be a sequence of positive linear operators from $C[a, b]$ into itself.

(b) Let $(\Omega, \mathcal{B}, P)$ be a probability space, and let $X(x, \omega)$ be a stochastic process from $[a, b] \times (\Omega, \mathcal{B}, P)$ into $\mathbb{R}$ such that $X(\cdot, \omega) \in C^m[a, b]$ for each $\omega \in \Omega$, and that $X^{(k)}(x, \cdot)$ is a measurable for all $k = 0, 1, ..., m$ and for each $x \in [a, b]$.

(c) Define the induced sequence (M_n) of positive linear operators on stochastic processes as follows:

$$M_n(X)(x, \omega) := L_n\left(X(\cdot, \omega); x\right) \quad \text{for} \quad \omega \in \Omega \quad \text{and} \quad x \in [a, b].$$

G.A. Anastassiou and O. Duman: Towards Intelligent Modeling, ISRL 14, pp. 131–142.
springerlink.com   © Springer-Verlag Berlin Heidelberg 2011

(d) Consider the corresponding expectation operator defined by

$$(EX)(x) := \int_{\Omega} X(x, \omega) P(d\omega) \quad \text{for} \ x \in [a, b].$$

(e) Assume that $X^{(m)}(x, \omega)$ is continuous in $x \in [a, b]$, uniformly with respect to $\omega \in \Omega$; that is, for every $\varepsilon > 0$, there is a $\delta > 0$ such that if $|y - x| \leq \delta$ $(x, y \in [a, b])$, then

$$\left| X^{(m)}(x, \omega) - X^{(m)}(x, \omega) \right| < \varepsilon$$

is satisfied for every $\omega \in \Omega$. In this case, we write $X^{(m)} \in C^U[a, b]$.

(f) Let $q \in (1, \infty)$. Then suppose that

$$\left(E \left| X^{(k)} \right|^q \right)(x) < \infty$$

holds for every $x \in [a, b]$ and for every $k = 0, 1, ..., m$.

(g) Assume that

$$\left(E \left| X^{(k)} \right| \right)(x) < \infty$$

holds for every $x \in [a, b]$ and for all $k = 0, 1, ..., m$.

As usual, we consider the test functions

$$e_i(y) := y^i, \quad i = 0, 1, 2 \ \text{ and } \ y \in [a, b],$$

and the moment function

$$\varphi_x(y) := y - x, \quad x, y \in [a, b].$$

Now, the q-mean first modulus of continuity of X (see [11]) is defined by

$$\Omega_1(X; \rho)_{L_q} := \sup_{\substack{|x-y| \leq \rho \\ (x,y \in [a,b])}} \left(\int_{\Omega} |X(x, \omega) - X(y, \omega)|^q P(d\omega) \right)^{\frac{1}{q}}, \quad \rho > 0, \ q \geq 1.$$

$$(10.1)$$

Then, we first need the following lemma.

Lemma 10.1. *Let $A = [a_{jn}]$ be a non-negative regular summability matrix. If (δ_n) is a sequence of positive real numbers such that $st_A - \lim_n \delta_n = 0$, then we get*

$$st_A - \lim_n \Omega_1(X; \delta_n)_{L_q} = 0.$$

Proof. As in the (b), let $X(x, \omega)$ be a stochastic process from $[a, b] \times (\Omega, \mathcal{B}, P)$. Since $st_A - \lim_n \delta_n = 0$, we obtain, for any $\delta > 0$, that

$$\lim_j \sum_{n : \delta_n \geq \delta} a_{jn} = 0. \tag{10.2}$$

By the right-continuity of $\Omega_1(X; \cdot)_{L_q}$ at zero, we can write that, for a given $\varepsilon > 0$, there is a $\delta > 0$ such that $\Omega_1(X; \rho)_{L_q} < \varepsilon$ whenever $0 < \rho < \delta$, i.e., $\Omega_1(X; \rho)_{L_q} \geq \varepsilon$ yields that $\rho \geq \delta$. Now replacing ρ by δ_n, for every $\varepsilon > 0$, we observe that

$$\{n : \Omega_1(X; \delta_n)_{L_q} \geq \varepsilon\} \subseteq \{n : \delta_n \geq \delta\}. \tag{10.3}$$

So, from (10.3), we get, for each $j \in \mathbb{N}$, that

$$\sum_{n : \Omega_1(X; \delta_n)_{L_q} \geq \varepsilon} a_{jn} \leq \sum_{n : \delta_n \geq \delta} a_{jn}. \tag{10.4}$$

Then, letting $j \to \infty$ on the both sides of inequality (10.4) and considering (10.2) we immediately see, for every $\varepsilon > 0$, that

$$\lim_j \sum_{n : \Omega_1(X; \delta_n)_{L_q} \geq \varepsilon} a_{jn} = 0$$

which gives $st_A - \lim_n \Omega_1(X; \delta_n)_{L_q} = 0$. Thus, the proof is finished. ∎

With this terminology, we obtain the following theorem.

Theorem 10.2. *Let $A = [a_{jn}]$ be a non-negative regular summability matrix, and let (L_n), $X(x, \omega)$, (M_n) and E be the same as in $(a) - (d)$, respectively. Assume that conditions (e) and (f) hold for a fixed $m \in \mathbb{N}$ and a fixed $q \in (1, \infty)$. Assume further that*

$$st_A - \lim_n \|L_n(e_0) - e_0\| = 0 \tag{10.5}$$

and

$$st_A - \lim_n \left\| L_n \left(|\varphi_x|^{q(m+1)} \right) \right\| = 0. \tag{10.6}$$

Then we get, for all X satisfying (e) and (f),

$$st_A - \lim_n \|E(|M_n(X) - X|^q)\| = 0. \tag{10.7}$$

Proof. Let $m \in \mathbb{N}$ and $q \in (1, \infty)$ be fixed. Then, by Corollary 2.1 of [11], we can write that, for all X satisfying (e) and (f) and for every $n \in \mathbb{N}$,

$$\|E(|M_n(X) - X|)\|^{1/q} \leq \|E(|X|^q)\|^{\frac{1}{q}} \|L_n(e_0) - e_0\|$$

$$+ \sum_{k=1}^{m} \frac{1}{k!} \left\|E\left(\left|X^{(k)}\right|^q\right)\right\|^{\frac{1}{q}} \left\|L_n(\varphi_x^k)\right\|$$

$$+ \lambda_{m,q} \|L_n(e_0)\|^{1-\frac{1}{q}} \left\|L_n(|\varphi_x|^{q(m+1)})\right\|^{\frac{m}{q(m+1)}}$$

$$\times \left((q+1)^{\frac{m}{m+1}} \|L_n(e_0)\|^{\frac{1}{m+1}} + 1\right)^{\frac{1}{q}}$$

$$\times \Omega_1 \left(X^{(m)}; \frac{1}{(q+1)^{\frac{1}{q(m+1)}}} \left\|L_n(|\varphi_x|^{q(m+1)})\right\|^{\frac{1}{q(m+1)}}\right)_{L_q},$$

where Ω_1 is given by (10.1), and

$$\lambda_{m,q} := \frac{1}{(m-1)!(q+1)^{\frac{m}{q(m+1)}}} \left(\frac{2(q-1)}{qm-1}\right)^{1-\frac{1}{q}}.$$

On the other hand, using Hölder's inequality with $\alpha = \frac{q(m+1)}{q(m+1)-k}$ and $\beta = \frac{q(m+1)}{k}$, $\frac{1}{\alpha} + \frac{1}{\beta} = 1$, we get that

$$\left\|L_n(\varphi_x^k)\right\| \leq \|L_n(e_0)\|^{\frac{q(m+1)-k}{q(m+1)}} \left\|L_n(|\varphi_x|^{q(m+1)})\right\|^{\frac{k}{q(m+1)}}, \quad k = 1, 2, ..., m.$$

Then, combining the above inequalities, we immediately see that

$$\|E(|M_n(X) - X|)\|^{1/q} \leq B_{m,q} \left\{ \sum_{k=1}^{m} \|L_n(e_0)\|^{\frac{q(m+1)-k}{q(m+1)}} \left\|L_n(|\varphi_x|^{q(m+1)})\right\|^{\frac{k}{q(m+1)}} \right.$$

$$+ \|L_n(e_0) - e_0\|$$

$$+ \|L_n(e_0)\|^{1-\frac{m}{q(m+1)}} \left\|L_n(|\varphi_x|^{q(m+1)})\right\|^{\frac{m}{q(m+1)}}$$

$$\times \Omega_1 \left(X^{(m)}; \delta_n(m,q)\right)_{L_q}$$

$$+ \|L_n(e_0)\|^{1-\frac{1}{q}} \left\|L_n(|\varphi_x|^{q(m+1)})\right\|^{\frac{m}{q(m+1)}}$$

$$\left. \Omega_1 \left(X^{(m)}; \delta_n(m,q)\right)_{L_q} \right\},$$

where

$$B_{m,q} := \max \left\{ \lambda_{m,q}(q+1)^{\frac{m}{q(m+1)}}, \|E(|X|^q)\|^{\frac{1}{q}}, \frac{\|E(|X'|^q)\|^{\frac{1}{q}}}{1!}, ..., \frac{\left\|E\left(\left|X^{(m)}\right|^q\right)\right\|^{\frac{1}{q}}}{m!} \right\}$$

and

$$\delta_n(m,q) := \frac{1}{(q+1)^{\frac{1}{q(m+1)}}} \left\| L_n(|\varphi_x|^{q(m+1)}) \right\|^{\frac{1}{q(m+1)}}. \qquad (10.8)$$

The hypotheses gives that $B_{m,q}$ is a finite positive number for each fixed $m \in \mathbb{N}$ and $q \in (1, \infty)$. Above we used the fact that

$$|x + y|^p \le |x|^p + |y|^p \quad \text{for} \quad p \in (0, 1]. \qquad (10.9)$$

We also derive

$$\|E\left(|M_n(X) - X|\right)\|^{1/q} \le B_{m,q} \left\{ \sum_{k=1}^{m} \|L_n(e_0) - e_0\|^{\frac{q(m+1)-k}{q(m+1)}} \left\| L_n(|\varphi_x|^{q(m+1)}) \right\|^{\frac{k}{q(m+1)}} \right.$$

$$+ \sum_{k=1}^{m} \left\| L_n(|\varphi_x|^{q(m+1)}) \right\|^{\frac{k}{q(m+1)}} + \|L_n(e_0) - e_0\|$$

$$+ \|L_n(e_0) - e_0\|^{1 - \frac{m}{q(m+1)}} \left\| L_n(|\varphi_x|^{q(m+1)}) \right\|^{\frac{m}{q(m+1)}}$$

$$\times \Omega_1 \left(X^{(m)}; \delta_n(m,q) \right)_{L_q}$$

$$+ \|L_n(e_0) - e_0\|^{1 - \frac{1}{q}} \left\| L_n(|\varphi_x|^{q(m+1)}) \right\|^{\frac{m}{q(m+1)}}$$

$$\times \Omega_1 \left(X^{(m)}; \delta_n(m,q) \right)_{L_q}$$

$$+ 2 \left\| L_n(|\varphi_x|^{q(m+1)}) \right\|^{\frac{m}{q(m+1)}} \Omega_1 \left(X^{(m)}; \delta_n(m,q) \right)_{L_q} \right\}.$$

Now, given $\varepsilon > 0$, consider the following sets:

$$V(\varepsilon) = \left\{ n : \|E\left(|M_n(X) - X|\right)\| \ge \varepsilon^q \right\},$$

$$V_1(\varepsilon) = \left\{ n : \|L_n(e_0) - e_0\| \ge \frac{\varepsilon}{2(m+2)B_{m,q}} \right\},$$

$$V_2(\varepsilon) = \left\{ n : \begin{array}{l} \|L_n(e_0) - e_0\|^{1 - \frac{m}{q(m+1)}} \left\| L_n(|\varphi_x|^{q(m+1)}) \right\|^{\frac{m}{q(m+1)}} \\ \times \Omega_1 \left(X^{(m)}; \delta_n(m,q) \right)_{L_q} \end{array} \ge \frac{\varepsilon}{2(m+2)B_{m,q}} \right\},$$

$$V_3(\varepsilon) = \left\{ n : \begin{array}{l} \|L_n(e_0) - e_0\|^{1 - \frac{1}{q}} \left\| L_n(|\varphi_x|^{q(m+1)}) \right\|^{\frac{m}{q(m+1)}} \\ \times \Omega_1 \left(X^{(m)}; \delta_n(m,q) \right)_{L_q} \end{array} \ge \frac{\varepsilon}{2(m+2)B_{m,q}} \right\},$$

$$V_4(\varepsilon) = \left\{ n : \left\| L_n(|\varphi_x|^{q(m+1)}) \right\|^{\frac{m}{q(m+1)}} \Omega_1 \left(X^{(m)}; \delta_n(m,q) \right)_{L_q} \ge \frac{\varepsilon}{4(m+2)B_{m,q}} \right\},$$

$$Y_k(\varepsilon) = \left\{ n : \|L_n(e_0) - e_0\|^{\frac{q(m+1)-k}{q(m+1)}} \left\| L_n(|\varphi_x|^{q(m+1)}) \right\|^{\frac{k}{q(m+1)}} \ge \frac{\varepsilon}{2(m+2)B_{m,q}} \right\},$$

$$Z_k(\varepsilon) = \left\{ n : \left\| L_n(|\varphi_x|^{q(m+1)}) \right\|^{\frac{k}{q(m+1)}} \ge \frac{\varepsilon}{2(m+2)B_{m,q}} \right\},$$

where $k = 1, 2, ..., m$. So, we have

$$V(\varepsilon) \subseteq \left(\bigcup_{i=1}^{4} V_i(\varepsilon)\right) \cup \left(\bigcup_{k=1}^{m} Y_k(\varepsilon)\right) \cup \left(\bigcup_{k=1}^{m} Z_k(\varepsilon)\right),$$

which implies

$$\delta_A(V(\varepsilon)) \leq \sum_{i=1}^{4} \delta_A(V_i(\varepsilon)) + \sum_{k=1}^{m} \delta_A(Y_k(\varepsilon)) + \sum_{k=1}^{m} \delta_A(Z_k(\varepsilon)). \qquad (10.10)$$

By (10.5), (10.6) and (10.8), one can show that, for each fixed $m \in \mathbb{N}$ and $q \in (1, \infty)$ and for every $k = 1, 2, ..., m$,

$$st_A - \lim_n \|L_n(e_0) - e_0\|^{\frac{q(m+1)-k}{q(m+1)}} = 0, \qquad (10.11)$$

$$st_A - \lim_n \left\|L_n(|\varphi_x|^{q(m+1)})\right\|^{\frac{k}{q(m+1)}} = 0, \qquad (10.12)$$

and

$$st_A - \lim_n \delta_n(m, q) = 0.$$

So, by Lemma 10.1, we get

$$st_A - \lim_n \Omega_1\left(X^{(m)}; \delta_n(m, q)\right)_{L_q} = 0. \qquad (10.13)$$

Now define

$$u_n := u_n(m, q, k) = \|L_n(e_0) - e_0\|^{\frac{q(m+1)-k}{q(m+1)}},$$

$$v_n := v_n(m, q, k) = \left\|L_n(|\varphi_x|^{q(m+1)})\right\|^{\frac{k}{q(m+1)}},$$

$$z_n := z_n(m, q) = \Omega_1\left(X^{(m)}; \delta_n(m, q)\right)_{L_q}.$$

Then, for every $\varepsilon > 0$, since

$$\{n : u_n v_n z_n \geq \varepsilon\} \subseteq \{n : u_n \geq \sqrt[3]{\varepsilon}\} \cup \{n : v_n \geq \sqrt[3]{\varepsilon}\} \cup \{n : z_n \geq \sqrt[3]{\varepsilon}\},$$

we observe that, for every $j \in \mathbb{N}$,

$$\sum_{n : u_n v_n z_n \geq \varepsilon} a_{jn} \leq \sum_{n : u_n \geq \sqrt[3]{\varepsilon}} a_{jn} + \sum_{n : v_n \geq \sqrt[3]{\varepsilon}} a_{jn} + \sum_{n : z_n \geq \sqrt[3]{\varepsilon}} a_{jn}.$$

If we take limit as $j \to \infty$ on the last inequality, and also consider (10.11), (10.12), (10.13), we immediately derive that

$$\lim_j \sum_{n : u_n v_n z_n \geq \varepsilon} a_{jn} = 0,$$

which gives

$$st_A - \lim_n u_n v_n z_n = 0. \tag{10.14}$$

Then, from (10.11)−(10.14), we get, for every $\varepsilon > 0$, that

$$\begin{aligned}
\delta_A(V_i(\varepsilon)) &= 0, \quad i = 1, 2, 3, 4, \\
\delta_A(Y_k(\varepsilon)) &= \delta_A(Z_k(\varepsilon)) = 0, \quad k = 1, 2, ..., m.
\end{aligned} \tag{10.15}$$

Thus, combining (10.10) with (10.15), we deduce that

$$\delta_A(V(\varepsilon)) = 0 \ \text{ for every } \ \varepsilon > 0,$$

which implies (10.7). The proof is done.   ∎

We also get the next result.

Theorem 10.3. *Let $A = [a_{jn}]$ be a non-negative regular summability matrix, and let (L_n), $X(x, \omega)$, (M_n) and E be the same as in $(a) - (d)$, respectively. Assume that conditions (e) and (g) hold for a fixed $m \in \mathbb{N}$. Assume further that*

$$st_A - \lim_n \|L_n(e_0) - e_0\| = 0 \tag{10.16}$$

and

$$st_A - \lim_n \left\| L_n\left(|\varphi_x|^{m+1}\right) \right\| = 0. \tag{10.17}$$

Then we get, for all X satisfying (e) and (g),

$$st_A - \lim_n \|E\left(|M_n(X) - X|\right)\| = 0. \tag{10.18}$$

Proof. Let $m \in \mathbb{N}$ be fixed. Then, it follows from Corollary 2.2 of [11] that, for all X satisfying (e) and (g) and for every $n \in \mathbb{N}$,

$$\begin{aligned}
\|E\left(|M_n(X) - X|\right)\| \leq \ & \|E\left(|X|\right)\| \, \|L_n(e_0) - e_0\| \\
& + \sum_{k=1}^m \frac{1}{k!} \left\| E\left(\left|X^{(k)}\right|\right) \right\| \, \|L_n(\varphi_x^k)\| \\
& + \frac{1}{m!} \left\| (L_n(e_0))^{\frac{1}{m+1}} + \frac{1}{m+1} \right\| \left\| L_n(|\varphi_x|^{m+1}) \right\|^{\frac{m}{m+1}} \\
& \times \Omega_1 \left(X^{(m)}; \left\| L_n(|\varphi_x|^{m+1}) \right\|^{\frac{1}{m+1}} \right)_{L_1},
\end{aligned}$$

where

$$\Omega_1(X; \rho)_{L_1} := \sup_{\substack{|x-y|\leq\rho \\ (x,y\in[a,b])}} \left(\int_\Omega |X(x, \omega) - X(y, \omega)| \, P(d\omega) \right), \quad \rho > 0.$$

Now applying Hölder's inequality with $\alpha = \frac{m+1}{m+1-k}$ and $\beta = \frac{m+1}{k}$, $\frac{1}{\alpha} + \frac{1}{\beta} = 1$, we see that

$$\left\| L_n(\varphi_x^k) \right\| \leq \left\| L_n(e_0) \right\|^{\frac{m+1-k}{m+1}} \left\| L_n(|\varphi_x|^{m+1}) \right\|^{\frac{k}{m+1}}, \quad k = 1, 2, ..., m.$$

Then, combining the above inequalities, we get

$$\left\| E\left(|M_n(X) - X| \right) \right\| \leq \mu_m \left\{ \sum_{k=1}^{m} \left\| L_n(e_0) \right\|^{\frac{m+1-k}{m+1}} \left\| L_n(|\varphi_x|^{m+1}) \right\|^{\frac{k}{m+1}} \right.$$
$$+ \left\| L_n(e_0) - e_0 \right\|$$
$$+ \left\| L_n(e_0) \right\|^{\frac{1}{m+1}} \left\| L_n(|\varphi_x|^{m+1}) \right\|^{\frac{m}{m+1}}$$
$$\times \Omega_1 \left(X^{(m)}; \rho_n(m) \right)_{L_1}$$
$$\left. + \left\| L_n(|\varphi_x|^{m+1}) \right\|^{\frac{m}{m+1}} \Omega_1 \left(X^{(m)}; \rho_n(m) \right)_{L_1} \right\},$$

where

$$\mu_m := \max \left\{ \frac{1}{m!}, \left\| E\left(|X| \right) \right\|, \frac{\left\| E\left(|X'| \right) \right\|}{1!}, \frac{\left\| E\left(|X''| \right) \right\|}{2!}, ..., \frac{\left\| E\left(|X^{(m)}| \right) \right\|}{m!} \right\}$$

and

$$\rho_n(m) := \left\| L_n(|\varphi_x|^{m+1}) \right\|^{\frac{1}{m+1}}. \tag{10.19}$$

Notice that the constant μ_m is a finite positive number for each fixed $m \in \mathbb{N}$. Also, using (10.9) we observe that

$$\left\| E\left(|M_n(X) - X| \right) \right\| \leq \mu_m \left\{ \sum_{k=1}^{m} \left\| L_n(e_0) - e_0 \right\|^{\frac{m+1-k}{m+1}} \left\| L_n(|\varphi_x|^{m+1}) \right\|^{\frac{k}{m+1}} \right.$$
$$+ \sum_{k=1}^{m} \left\| L_n(|\varphi_x|^{m+1}) \right\|^{\frac{k}{m+1}} + \left\| L_n(e_0) - e_0 \right\|$$
$$+ \left\| L_n(e_0) - e_0 \right\|^{\frac{1}{m+1}} \left\| L_n(|\varphi_x|^{m+1}) \right\|^{\frac{m}{m+1}}$$
$$\times \Omega_1 \left(X^{(m)}; \rho_n(m) \right)_{L_1}$$
$$\left. + 2 \left\| L_n(|\varphi_x|^{m+1}) \right\|^{\frac{m}{m+1}} \Omega_1 \left(X^{(m)}; \rho_n(m) \right)_{L_1} \right\}.$$

Now, as in the proof of Theorem 10.2, given $\varepsilon > 0$, consider the following sets:

$$K(\varepsilon) = \{n : \|E\left(|M_n(X) - X|\right)\| \geq \varepsilon\},$$

$$K_1(\varepsilon) = \left\{n : \|L_n(e_0) - e_0\| \geq \frac{\varepsilon}{(2m+3)\mu_m}\right\},$$

$$K_2(\varepsilon) = \left\{n : \begin{array}{c} \|L_n(e_0) - e_0\|^{\frac{1}{m+1}} \left\||L_n(|\varphi_x|^{m+1})\right\|^{\frac{m}{m+1}} \\ \times \Omega_1\left(X^{(m)}; \rho_n(m)\right)_{L_1} \end{array} \geq \frac{\varepsilon}{(2m+3)\mu_m}\right\},$$

$$K_3(\varepsilon) = \left\{n : \left\|L_n(|\varphi_x|^{m+1})\right\|^{\frac{m}{m+1}} \Omega_1\left(X^{(m)}; \rho_n(m)\right)_{L_1} \geq \frac{\varepsilon}{2(2m+3)\mu_m}\right\},$$

$$L_k(\varepsilon) = \left\{n : \|L_n(e_0) - e_0\|^{\frac{m+1-k}{m+1}} \left\|L_n(|\varphi_x|^{m+1})\right\|^{\frac{k}{m+1}} \geq \frac{\varepsilon}{(2m+3)\mu_m}\right\},$$

$$M_k(\varepsilon) = \left\{n : \left\|L_n(|\varphi_x|^{m+1})\right\|^{\frac{k}{m+1}} \geq \frac{\varepsilon}{(2m+3)\mu_m}\right\},$$

where $k = 1, 2, ..., m$. So, for every $\varepsilon > 0$, we have

$$K(\varepsilon) \subseteq \left(\bigcup_{i=1}^{3} K_i(\varepsilon)\right) \cup \left(\bigcup_{k=1}^{m} L_k(\varepsilon)\right) \cup \left(\bigcup_{k=1}^{m} M_k(\varepsilon)\right),$$

which gives

$$\delta_A(K(\varepsilon)) \leq \sum_{i=1}^{3} \delta_A(K_i(\varepsilon)) + \sum_{k=1}^{m} \delta_A(L_k(\varepsilon)) + \sum_{k=1}^{m} \delta_A(M_k(\varepsilon)). \qquad (10.20)$$

Observe that the hypotheses (10.16) and (10.17) implies, for every $k = 1, 2, ..., m$,

$$st_A - \lim_n \|L_n(e_0) - e_0\|^{\frac{m+1-k}{m+1}} = 0,$$

$$st_A - \lim_n \left\|L_n(|\varphi_x|^{m+1})\right\|^{\frac{k}{m+1}} = 0$$

and

$$st_A - \lim_n \rho_n(m) = 0.$$

Then, by considering Lemma 10.1 and by using a similar technique to the proof of Theorem 10.2, we can write that

$$st_A - \lim_n \|L_n(e_0) - e_0\|^{\frac{m+1-k}{m+1}} \left\|L_n(|\varphi_x|^{m+1})\right\|^{\frac{k}{m+1}} = 0,$$

$$st_A - \lim_n \|L_n(e_0) - e_0\|^{\frac{1}{m+1}} \left\|L_n(|\varphi_x|^{m+1})\right\|^{\frac{m}{m+1}} \Omega_1\left(X^{(m)}; \rho_n(m)\right)_{L_1} = 0$$

and

$$st_A - \lim_n \left\| L_n(|\varphi_x|^{m+1}) \right\|^{\frac{m}{m+1}} \Omega_1 \left(X^{(m)}; \rho_n(m) \right)_{L_1} = 0.$$

So, we obtain the following results:

$$\delta_A(K_i(\varepsilon)) = 0, \quad i = 1, 2, 3,$$
$$\delta_A(L_k(\varepsilon)) = \delta_A(M_k(\varepsilon)) = 0, \quad k = 1, 2, ..., m. \tag{10.21}$$

Hence, combining (10.20) with (10.21), we deduce that

$$\delta_A(K(\varepsilon)) = 0 \text{ for every } \varepsilon > 0,$$

which implies (10.18). The proof is completed.   ∎

10.2 Conclusions

In this section, we present some consequences of Theorems 10.2 and 10.3. We also introduce a sequence of positive linear operators which satisfies our results but not their classical cases.

If we replace the matrix $A = [a_{jn}]$ by the identity matrix in Theorems 10.2 and 10.3, then we immediately get the following results introduced in [11].

Corollary 10.4 (see [11]). Let (L_n), $X(x, \omega)$, (M_n) and E be the same as in $(a) - (d)$, respectively. Assume that conditions (e) and (g) hold for a fixed $m \in \mathbb{N}$ and a fixed $q \in (1, \infty)$. Assume further that the sequences $(L_n(e_0))$ and $\left(L_n \left(|\varphi_x|^{q(m+1)} \right) \right)$ are uniformly convergent to e_0 and 0 on $[a, b]$, respectively. Then, for all X satisfying (e) and (f), the sequence $(E(|M_n(X) - X|^q))$ is uniformly convergent to 0 on $[a, b]$.

Corollary 10.5 (see [11]). Let (L_n), $X(x, \omega)$, (M_n) and E be the same as in $(a)-(d)$, respectively. Assume that conditions (e) and (g) hold for a fixed $m \in \mathbb{N}$. Assume further that the sequences $(L_n(e_0))$ and $\left(L_n \left(|\varphi_x|^{m+1} \right) \right)$ are uniformly convergent to e_0 and 0 on $[a, b]$, respectively. Then, for all X satisfying (e) and (g), the sequence $(E(|M_n(X) - X|))$ is uniformly convergent to 0 on $[a, b]$.

Now considering Corollaries 2.3 and 2.4 of [11] and also using the similar techniques as in Theorems 10.2 and 10.3, one can obtain the following results in the case of $m = 0$.

Corollary 10.6. Let $A = [a_{jn}]$ be a non-negative regular summability matrix, and let (L_n), $X(x, \omega)$, (M_n) and E be the same as in $(a) - (d)$, respectively. Assume that conditions (e) and (f) hold for $m = 0$ and a fixed $q \in (1, \infty)$. Assume further that

$$st_A - \lim_n \|L_n(e_0) - e_0\| = 0$$

and

$$st_A - \lim_n \|L_n (|\varphi_x|^q)\| = 0.$$

Then, we get, for all X satisfying (e) and (f),

$$st_A - \lim_n \|E (|M_n(X) - X|^q)\| = 0.$$

Corollary 10.7. *Let $A = [a_{jn}]$ be a non-negative regular summability matrix, and let (L_n), $X(x,\omega)$, (M_n) and E be the same as in $(a) - (d)$, respectively. Assume that conditions (e) and (g) hold for $m = 0$. Assume further that*

$$st_A - \lim_n \|L_n(e_i) - e_i\| = 0, \quad i = 0, 1, 2.$$

Then we get, for all X satisfying (e) and (g),

$$st_A - \lim_n \|E (|M_n(X) - X|)\| = 0.$$

Also, taking $A = I$, the identity matrix, in Corollaries 10.6 and 10.7, we get the following results.

Corollary 10.8 (see [11]). *Let (L_n), $X(x,\omega)$, (M_n) and E be the same as in $(a) - (d)$, respectively. Assume that conditions (e) and (f) hold for $m = 0$ and a fixed $q \in (1, \infty)$. Assume further that the sequences $(L_n(e_0))$ and $(L_n (|\varphi_x|^q))$ are uniformly convergent to e_0 and 0 on $[a, b]$, respectively. Then, for all X satisfying (e) and (f), the sequence $(E (|M_n(X) - X|^q))$ is uniformly convergent to 0 on $[a, b]$.*

Corollary 10.9 (see [11]). *Let (L_n), $X(x,\omega)$, (M_n) and E be the same as in $(a) - (d)$, respectively. Assume that conditions (e) and (g) hold for $m = 0$. Assume further that, for each $i = 0, 1, 2$, $(L_n(e_i))$ is uniformly convergent to e_i on $[a, b]$. Then, for all X satisfying (e) and (g), the sequence $(E (|M_n(X) - X|))$ is uniformly convergent to 0 on $[a, b]$.*

Finally, we give an example as follows. Consider the classical Bernstein polynomials given by

$$B_n(f; x) = \sum_{k=0}^{n} f \left(\frac{k}{n} \right) \binom{n}{k} x^k (1 - x)^{n-k},$$

where $x \in [0, 1]$, $f \in C[0, 1]$, $n \in \mathbb{N}$. Also define $M_n(X)$ by

$$M_n(X)(x, \omega) : = B_n(X(\cdot, \omega); x)$$
$$= \sum_{k=0}^{n} X \left(\frac{k}{n}, \omega \right) \binom{n}{k} x^k (1 - x)^{n-k}, \tag{10.22}$$

where $x \in [0, 1]$, $\omega \in \Omega$, $n \in \mathbb{N}$, and X is a stochastic process satisfying (e) and (g) for $m = 0$. In this case, by Corollary 10.9 (see also [11]) we have, for all such X's, the sequence $(E(|M_n(X) - X|))$ is uniformly convergent to 0 on the interval $[0, 1]$.

Now take $A = C_1$, the Cesáro matrix, and define a sequence (u_n) by

$$u_n := \begin{cases} 1, & \text{if } n = k^2, \ k \in \mathbb{N} \\ 0, & \text{otherwise.} \end{cases}$$

Then we see that (u_n) is non-convergent (in the usual sense) but $st_{C_1} - \lim_n u_n = st - \lim_n u_n = 0$. Now define the following positive linear operators

$$L_n(f; x) := (1 + u_n)B_n(f; x), \qquad (10.23)$$

where B_n are the Bernstein polynomials. Also define $M_n^*(X)$ by

$$M_n^*(X)(x, \omega) := (1 + u_n)M_n(X), \qquad (10.24)$$

where $M_n(X)$ is given by (10.22). Since (u_n) is non-convergent, the sequence $(E(|M_n^*(X) - X|))$ is not uniformly convergent to 0 on $[0, 1]$. So, Corollary 10.9 fails for the operators L_n defined by (10.23) and the induced sequence (M_n^*) defined by (10.24). However, we see that these sequences satisfy all assumptions in Corollary 10.7 whenever $A = C_1$ and $m = 0$. So, we obtain that if X satisfies (e) and (g) for $m = 0$, then

$$st - \lim_n \|E(|M_n^*(X) - X|)\| = 0.$$

This example demonstrates that the statistical approximation results for stochastic process are stronger than their classical cases introduced in [11].

11

Statistical Approximation Theory for Multivariate Stochastic Processes

In this chapter, we obtain some Korovkin-type approximation theorems for multivariate stochastic processes with the help of the concept of A-statistical convergence. A non-trivial example showing the importance of this method of approximation is also introduced. This chapter relies on [26].

11.1 Statistical Korovkin-Type Results for Multivariate Stochastic Processes

Let $m \in \mathbb{N}_0$, the set of all non-negative integers, and let $\mathbf{Q}$ be a compact and convex subset of $\mathbb{R}^k$, $k > 1$. Then, as usual, by $C^m(\mathbf{Q})$ we denote the space of all m-times continuously differentiable functions on $\mathbf{Q}$ endowed with the sup-norm $\|\cdot\|$. Then, throughout this section we consider the following concepts and assumptions (*cf.* [12]):

(a) Let (L_n) be sequence of positive linear operators from $C(\mathbf{Q})$ into itself.

(b) Let $(\Omega, \mathcal{B}, P)$ be a probability space, and let $X(\mathbf{x}, \omega)$ be a multivariate stochastic process from $\mathbf{Q} \times (\Omega, \mathcal{B}, P)$ into $\mathbb{R}$ such that $X(\cdot, \omega) \in C^m(\mathbf{Q})$ for each $\omega \in \Omega$, and that $X_{\boldsymbol{\alpha}}(\mathbf{x}, \cdot) := \dfrac{\partial^{\boldsymbol{\alpha}} X}{\partial t^{\boldsymbol{\alpha}}}(\mathbf{x}, \cdot)$ is a measurable for each $\mathbf{x} \in \mathbf{Q}$ and for all $\boldsymbol{\alpha} = (\alpha_1, ..., \alpha_k)$, $\alpha_i \in \mathbb{N}_0$, $i = 1, .., k$, $0 \leq |\boldsymbol{\alpha}| \leq m$ with $|\boldsymbol{\alpha}| := \sum_{i=1}^{k} \alpha_i$.

G.A. Anastassiou and O. Duman: Towards Intelligent Modeling, ISRL 14, pp. 143–155.
springerlink.com   © Springer-Verlag Berlin Heidelberg 2011

(c) Consider the induced sequence (M_n) of positive linear operators on multivariate stochastic processes defined by

$$M_n(X)(\mathbf{x}, \omega) := L_n\left(X(\cdot, \omega); \mathbf{x}\right) \quad \text{for} \quad \omega \in \Omega \text{ and } \mathbf{x} \in \mathbf{Q},$$

where $M_n(X)$ is assumed to be measurable in $\omega \in \Omega$.

(d) Define the corresponding expectation operator as follows:

$$(EX)(\mathbf{x}) := \int_{\Omega} X(\mathbf{x}, \omega) P(d\omega) \quad \text{for} \quad \mathbf{x} \in [a, b].$$

(e) Suppose that $X_{\boldsymbol{\alpha}}(\mathbf{x}, \omega)$, $|\boldsymbol{\alpha}| = m$, is continuous in $\mathbf{x} \in \mathbf{Q}$, uniformly with respect to $\omega \in \Omega$; that is, for every $\varepsilon > 0$, there exists a $\delta > 0$ such that whenever $\|\mathbf{y} - \mathbf{x}\|_{\ell_1} \leq \delta$ $(\mathbf{x}, \mathbf{y} \in \mathbf{Q})$, then

$$|X_{\boldsymbol{\alpha}}(\mathbf{x}, \omega) - X_{\boldsymbol{\alpha}}(\mathbf{y}, \omega)| < \varepsilon$$

holds for all $\omega \in \Omega$. Then, we write $X_{\boldsymbol{\alpha}} \in C_{\mathbb{R}}^U(\mathbf{Q})$.

(f) Let $q \in (1, \infty)$. Assume that

$$\left(E |X_{\boldsymbol{\alpha}}|^q\right)(\mathbf{x}) < \infty$$

holds for every $\mathbf{x} \in \mathbf{Q}$ and for all $\boldsymbol{\alpha} = (\alpha_1, ..., \alpha_k)$, $\alpha_i \in \mathbb{N}_0$, $i = 1, .., k$, $0 \leq |\boldsymbol{\alpha}| \leq m$.

(g) Assume that

$$\left(E |X_{\boldsymbol{\alpha}}|\right)(\mathbf{x}) < \infty$$

holds for every $\mathbf{x} \in \mathbf{Q}$ and for all $\boldsymbol{\alpha} = (\alpha_1, ..., \alpha_k)$, $\alpha_i \in \mathbb{N}_0$, $i = 1, .., k$, $0 \leq |\boldsymbol{\alpha}| \leq m$.

In this section, we consider the test function

$$e_0(\mathbf{y}) := 1 \quad \text{for} \quad \mathbf{y} \in \mathbf{Q},$$

and also the moment function

$$\varphi_{\mathbf{x}}(\mathbf{y}) := \|\mathbf{y} - \mathbf{x}\|_{\ell_1} \quad \text{for} \quad \mathbf{x}, \mathbf{y} \in \mathbf{Q}.$$

If $\mathbf{Q}$ is a compact and convex subset of $\mathbb{R}^k$ and let $X(\mathbf{x}, \omega)$ be a multivariate stochastic process from $\mathbf{Q} \times (\Omega, \mathcal{B}, P)$ into $\mathbb{R}$, where $(\Omega, \mathcal{B}, P)$ is a probability space. Then, in [12], the q-mean multivariate first modulus of continuity of X, denoted by $\Omega_1(X; \rho)_{L_q}$, is defined by

$$\Omega_1(X; \rho)_{L_q} := \sup_{\substack{\|\mathbf{x} - \mathbf{y}\|_{\ell_1} \leq \rho \\ (\mathbf{x}, \mathbf{y} \in \mathbf{Q})}} \left(\int_{\Omega} |X(\mathbf{x}, \omega) - X(\mathbf{y}, \omega)|^q P(d\omega)\right)^{1/q}, \quad \rho > 0, \ q \geq 1.$$

$$(11.1)$$

The next proposition (see [12]) gives the main properties of $\Omega_1(X; \rho)_{L_q}$.

Proposition 11.1 (see [12]). *Let $X(\mathbf{x}, \omega)$ and $Y(\mathbf{x}, \omega)$ be two multivariate stochastic processes from $\mathbf{Q} \times (\Omega, \mathcal{B}, P)$ into $\mathbb{R}$. Then we get:*

(i) $\Omega_1 (X; \rho)_{L_q}$ is non-negative and non-decreasing in $\rho > 0$.

(ii) $\lim_{\rho \searrow 0} \Omega_1 (X; \rho)_{L_q} = \Omega_1 (X; 0)_{L_q} = 0$ if and only if $X \in C_{\mathbb{R}}^{U_q}(\mathbf{Q})$, that is, for every $\varepsilon > 0$, there exists a $\delta > 0$ such that whenever $\|\mathbf{y} - \mathbf{x}\|_{\ell_1} \le \delta$ $(\mathbf{x}, \mathbf{y} \in \mathbf{Q})$, then the inequality

$$\int_{\Omega} |X_\alpha(\mathbf{x}, \omega) - X_\alpha(\mathbf{y}, \omega)|^q \, P(d\omega) < \varepsilon$$

holds.

(iii) $\Omega_1 (X; \rho_1 + \rho_2)_{L_q} \le \Omega_1 (X; \rho_1)_{L_q} + \Omega_1 (X; \rho_2)_{L_q}$ for any $\rho_1, \rho_2 > 0$.

(iv) $\Omega_1 (X; n\rho)_{L_q} \le n \, \Omega_1 (X; \rho)_{L_q}$ for any $\rho > 0$ and $n \in \mathbb{N}$.

(v) $\Omega_1 (X; \lambda\rho)_{L_q} \le \lceil \lambda \rceil \Omega_1 (X; \rho)_{L_q} \le (\lambda + 1)\Omega_1 (X; \rho)_{L_q}$ for any $\rho, \lambda > 0$, where $\lceil \lambda \rceil$ is ceiling of the number, that is the smallest integer greater equal the number.

(vi) $\Omega_1 (X + Y; \rho)_{L_q} \le \Omega_1 (X; \rho)_{L_q} + \Omega_1 (Y; \rho)_{L_q}$ for any $\rho > 0$.

(vii) $\Omega_1 (X; \cdot)_{L_q}$ is continuous on $\mathbb{R}_+$ for $X \in C_{\mathbb{R}}^{U_q}(\mathbf{Q})$.

We also need the next lemma.

Lemma 11.2. *Let $A = [a_{jn}]$ be a non-negative regular summability matrix. If (δ_n) is a sequence of positive real numbers such that $st_A - \lim_n \delta_n = 0$, then, for any multivariate stochastic process X as in (b) and (e), we get*

$$st_A - \lim_n \Omega_1 (X; \delta_n)_{L_q} = 0.$$

Proof. Since $st_A - \lim_n \delta_n = 0$, we obtain, for any $\delta > 0$, that

$$\lim_j \sum_{n:\delta_n \ge \delta} a_{jn} = 0. \tag{11.2}$$

By (e) we get $X_\alpha \in C_{\mathbb{R}}^{U}(\mathbf{Q})$. This implies that, for a given $\varepsilon > 0$, there exists a $\delta > 0$ such that whenever $\|\mathbf{y} - \mathbf{x}\|_{\ell_1} \le \delta$ $(\mathbf{x}, \mathbf{y} \in \mathbf{Q})$,

$$|X_\alpha(\mathbf{x}, \omega) - X_\alpha(\mathbf{y}, \omega)| < \varepsilon.$$

So, we observe that

$$\left(\int_{\Omega} |X_\alpha(\mathbf{x}, \omega) - X_\alpha(\mathbf{y}, \omega)|^q \, P(d\omega) \right)^{1/q} < \varepsilon \quad \text{for} \quad q \ge 1.$$

The last inequality gives that $\Omega_1(X;\rho)_{L_q} < \varepsilon$ whenever $0 < \rho < \delta$. Then, we can write that $\Omega_1(X;\rho)_{L_q} \geq \varepsilon$ yields $\rho \geq \delta$. Now replacing ρ by δ_n, for every $\varepsilon > 0$, we obtain that

$$\{n \in \mathbb{N} : \Omega_1(X;\delta_n)_{L_q} \geq \varepsilon\} \subseteq \{n \in \mathbb{N} : \delta_n \geq \delta\}. \tag{11.3}$$

So, from (11.3), we have for each $j \in \mathbb{N}$, that

$$\sum_{n:\Omega_1(X;\delta_n)_{L_q}\geq\varepsilon} a_{jn} \leq \sum_{n:\delta_n\geq\delta} a_{jn}. \tag{11.4}$$

Then, letting $j \to \infty$ on the both sides of inequality (11.4) and using (11.2) we immediately see, for every $\varepsilon > 0$, that

$$\lim_j \sum_{n:\Omega_1(X;\delta_n)_{L_q}\geq\varepsilon} a_{jn} = 0,$$

which implies that $st_A - \lim_n \Omega_1(X;\delta_n)_{L_q} = 0$. Thus, the proof is finished. $\blacksquare$

Therefore, we are ready to give the first approximation result.

Theorem 11.3. *Let $A = [a_{jn}]$ be a non-negative regular summability matrix, and let (L_n), $X(x,\omega)$, (M_n) and E be the same as in $(a) - (d)$, respectively. Assume that conditions (e) and (f) hold for a fixed $m \in \mathbb{N}$ and a fixed $q \in (1,\infty)$. Assume further that*

$$st_A - \lim_n \|L_n(e_0) - e_0\| = 0 \tag{11.5}$$

and

$$st_A - \lim_n \left\|L_n\left((\varphi_{\mathbf{x}})^{q(m+1)}\right)\right\| = 0. \tag{11.6}$$

Then we get, for all X satisfying (e) and (f),

$$st_A - \lim_n \|E(|M_n(X) - X|^q)\| = 0. \tag{11.7}$$

Proof. Let $m \in \mathbb{N}$ and $q \in (1,\infty)$ be fixed. Then, by Theorem 6 of [12], we can write that, for all X satisfying (e) and (f) and for every $n \in \mathbb{N}$,

$$\|E(|M_n(X) - X|^q)\|^{1/q} \leq \|E(|X|^q)\|^{\frac{1}{q}}\|L_n(e_0) - e_0\|$$

$$+ \left(\sum_{i=1}^m \lambda_j(k,q) \left\{\left(\sum_{|\alpha|=i} \mu_i(\alpha,q)\right) \left\|L_n\left((\varphi_{\mathbf{x}})^{qi}\right)\right\|\right\}^{\frac{1}{q}}\right) \|L_n(e_0)\|^{1-\frac{1}{q}}$$

$$+ \xi(k,m,q) \|L_n(e_0)\|^{1-\frac{1}{q}} \left\|\frac{(L_n(e_0))^{\frac{1}{m+1}}}{m!} + \tau(m,q)\right\|^{\frac{1}{q}}$$

$$\times \left\|L_n\left((\varphi_{\mathbf{x}})^{q(m+1)}\right)\right\|^{\frac{m}{q(m+1)}} \left\{\max_{\alpha:|\alpha|=m} \Omega_1\left(X_\alpha; \left\|L_n\left((\varphi_{\mathbf{x}})^{q(m+1)}\right)\right\|^{\frac{1}{q(m+1)}}\right)_{L_q}\right\},$$

where

$$\lambda_i(k,q) := \frac{k^{i\left(1-\frac{1}{q}\right)}}{(i!)^{1-\frac{1}{q}}}, \quad \mu_i(\alpha,q) := \frac{\|E\left(|X_\alpha|^q\right)\|}{i!} \quad \text{for} \ \ i = 1,...,m$$

and

$$\xi(k,m,q) := \left(\frac{2k^m}{m!}\right)^{1-\frac{1}{q}}, \quad \tau(m,q) := \frac{1}{(q+1)...(q+m)}.$$

By (11.5), the sequence $(\|L_n(e_0)\|)$ is A-statistically bounded, i.e., there exists a subsequence K of $\mathbb{N}$ with A-density one such that, for every $n \in K$, $\|L_n(e_0)\| \leq M$ holds for some $M > 0$. Then, the above inequality gives that, for every $n \in K$,

$$\|E\left(|M_n(X) - X|^q\right)\|^{1/q} \leq \|E\left(|X|^q\right)\|^{\frac{1}{q}} \|L_n(e_0) - e_0\|$$
$$+ \left(\sum_{i=1}^{m} \lambda_j(k,q) \left\{\left(\sum_{|\alpha|=i} \mu_i(\alpha,q)\right) \left\|L_n\left((\varphi_{\mathbf{x}})^{qi}\right)\right\|\right\}^{\frac{1}{q}}\right) M^{1-\frac{1}{q}}$$
$$+ \xi(k,m,q)M^{1-\frac{1}{q}} \left\|\frac{(L_n(e_0))^{\frac{1}{m+1}}}{m!} + \tau(m,q)\right\|^{\frac{1}{q}}$$
$$\times \left\|L_n\left((\varphi_{\mathbf{x}})^{q(m+1)}\right)\right\|^{\frac{m}{q(m+1)}} \left\{\max_{\alpha:|\alpha|=m} \Omega_1\left(X_\alpha; \left\|L_n\left((\varphi_{\mathbf{x}})^{q(m+1)}\right)\right\|^{\frac{1}{q(m+1)}}\right)_{L_q}\right\}.$$

On the other hand, using Hölder's inequality with $u = \frac{m+1}{m+1-i}$ and $v = \frac{m+1}{i}$, $\frac{1}{u} + \frac{1}{v} = 1$, we see that

$$\left\|L_n\left(\varphi_{\mathbf{x}}^{qi}\right)\right\| \leq \|L_n(e_0)\|^{\frac{m+1-i}{m+1}} \left\|L_n\left((\varphi_{\mathbf{x}})^{q(m+1)}\right)\right\|^{\frac{i}{m+1}}, \quad i = 1,2,...,m.$$

Then, combining the above inequalities, we immediately obtain, for every $n \in K$,

$$\|E\left(|M_n(X) - X|^q\right)\|^{1/q} \leq \|E\left(|X|^q\right)\|^{\frac{1}{q}} \|L_n(e_0) - e_0\|$$
$$+ \left(\sum_{i=1}^{m} \lambda_i(k,q) \left\{\left(\sum_{|\alpha|=i} \mu_i(\alpha,q)\right) \|L_n(e_0)\|^{\frac{m+1-i}{m+1}} \left\|L_n\left((\varphi_{\mathbf{x}})^{q(m+1)}\right)\right\|^{\frac{i}{m+1}}\right\}^{\frac{1}{q}}\right) M^{1-\frac{1}{q}}$$
$$+ \xi(k,m,q)M^{1-\frac{1}{q}} \left\|\frac{(L_n(e_0))^{\frac{1}{m+1}}}{m!} + \tau(q,m)\right\|^{\frac{1}{q}}$$
$$\times \left\|L_n\left((\varphi_{\mathbf{x}})^{q(m+1)}\right)\right\|^{\frac{m}{q(m+1)}} \left\{\max_{\alpha:|\alpha|=m} \Omega_1\left(X_\alpha; \left\|L_n\left((\varphi_{\mathbf{x}})^{q(m+1)}\right)\right\|^{\frac{1}{q(m+1)}}\right)_{L_q}\right\},$$

which implies that

$$
\begin{aligned}
\left\| E\left(|M_n(X) - X|^q\right) \right\|^{1/q} &\leq A(q)\left\| L_n(e_0) - e_0 \right\| \\
&+ B(k,m,q) \sum_{i=1}^{m} \left\| L_n\left((\varphi_{\mathbf{x}})^{q(m+1)}\right) \right\|^{\frac{i}{q(m+1)}} \\
&+ C(k,m,q) \left\| L_n\left((\varphi_{\mathbf{x}})^{q(m+1)}\right) \right\|^{\frac{m}{q(m+1)}} \\
&\times \left\{ \max_{\boldsymbol{\alpha}:|\boldsymbol{\alpha}|=m} \Omega_1\left(X_{\boldsymbol{\alpha}}; \left\| L_n\left((\varphi_{\mathbf{x}})^{q(m+1)}\right) \right\|^{\frac{1}{q(m+1)}}\right)_{L_q} \right\},
\end{aligned} \tag{11.8}
$$

where

$$
A(q) := \left\| E\left(|X|^q\right) \right\|^{\frac{1}{q}},
$$

$$
B(k,m,q) := M^{1-\frac{1}{q}} \max_{i=1,..,m}\left\{ \lambda_i(k,q)\left\{ \left(\sum_{|\boldsymbol{\alpha}|=i} \mu_i(\boldsymbol{\alpha},q)\right) M^{\frac{m+1-i}{m+1}} \right\}^{\frac{1}{q}} \right\},
$$

$$
C(k,m,q) := \xi(k,m,q) M^{1-\frac{1}{q}}\left(\frac{M^{\frac{1}{m+1}}}{m!} + \tau(q,m) \right)^{\frac{1}{q}}.
$$

The hypotheses gives that the constants $A(q)$, $B(k,m,q)$ and $C(k,m,q)$ are finite positive numbers for each fixed $k, m \in \mathbb{N}$ and $q \in (1,\infty)$. Now, given $\varepsilon > 0$, define the following sets:

$$
V(\varepsilon) = \left\{ n : \left\| E\left(|M_n(X) - X|^q\right) \right\| \geq \varepsilon^q \right\},
$$

$$
V_0(\varepsilon) = \left\{ n : \left\| L_n(e_0) - e_0 \right\| \geq \frac{\varepsilon}{(m+3)A(q)} \right\},
$$

$$
V_i(\varepsilon) = \left\{ n : \left\| L_n\left((\varphi_{\mathbf{x}})^{q(m+1)}\right) \right\|^{\frac{i}{q(m+1)}} \geq \frac{\varepsilon}{(m+3)B(k,m,q)} \right\}, \quad i = 1,2,...,m,
$$

$$
V_{m+1}(\varepsilon) = \left\{ n : \left\| L_n\left((\varphi_{\mathbf{x}})^{q(m+1)}\right) \right\|^{\frac{m}{q(m+1)}} \geq \sqrt{\frac{2\varepsilon}{(m+3)C(k,m,q)}} \right\},
$$

$$
V_{m+2}(\varepsilon) = \left\{ n : \max_{\boldsymbol{\alpha}:|\boldsymbol{\alpha}|=m} \Omega_1\left(X_{\boldsymbol{\alpha}}; \left\| L_n\left((\varphi_{\mathbf{x}})^{q(m+1)}\right) \right\|^{\frac{1}{q(m+1)}}\right)_{L_q} \geq \sqrt{\frac{2\varepsilon}{(m+3)C(k,m,q)}} \right\}.
$$

Hence, we obtain from (11.8) that

$$
V(\varepsilon) \cap K \subseteq \bigcup_{i=0}^{m+2} V_i(\varepsilon) \cap K,
$$

which implies

$$
\sum_{n \in V(\varepsilon) \cap K} a_{jn} \leq \sum_{i=0}^{m+2} \sum_{n \in V_i(\varepsilon) \cap K} a_{jn} \leq \sum_{i=0}^{m+2} \sum_{n \in V_i(\varepsilon)} a_{jn} \quad \text{for every } j \in \mathbb{N}. \tag{11.9}
$$

By (11.5), we derive

$$\lim_{j \to \infty} \sum_{n \in V_0(\varepsilon)} a_{jn} = 0. \tag{11.10}$$

Also, by (11.6), we see that

$$\lim_{j \to \infty} \sum_{n \in V_i(\varepsilon)} a_{jn} = 0 \text{ for each } i = 1, 2, ..., m. \tag{11.11}$$

Again from (11.6), we get

$$\lim_{j \to \infty} \sum_{n \in V_{m+1}(\varepsilon)} a_{jn} = 0. \tag{11.12}$$

Now, it follows from (11.6) and Lemma 11.2 that

$$\lim_{j \to \infty} \sum_{n \in V_{m+2}(\varepsilon)} a_{jn} = 0. \tag{11.13}$$

Hence, letting $j \to \infty$ in (11.9) and also combining (11.10) $-$ (11.13) we immediately obtain that

$$\lim_{j \to \infty} \sum_{n \in V(\varepsilon) \cap K} a_{jn} = 0. \tag{11.14}$$

Furthermore, we have

$$\sum_{n \in V(\varepsilon)} a_{jn} = \sum_{n \in V(\varepsilon) \cap K} a_{jn} + \sum_{n \in V(\varepsilon) \cap (\mathbb{N} \setminus K)} a_{jn} \leq \sum_{n \in V(\varepsilon) \cap K} a_{jn} + \sum_{n \in \mathbb{N} \setminus K} a_{jn}. \tag{11.15}$$

Since $\delta_A(K) = 1$, it is clear that $\delta_A(\mathbb{N} \setminus K) = 0$. So, taking limit as $j \to \infty$ in (11.15), and considering (11.14), we observe that

$$\lim_{j \to \infty} \sum_{n \in V(\varepsilon)} a_{jn} = 0,$$

which gives (11.7). Therefore, the proof is done.   ∎

We also get the following result.

Theorem 11.4. *Let $A = [a_{jn}]$ be a non-negative regular summability matrix, and let (L_n), $X(x, \omega)$, (M_n) and E be the same as in $(a) - (d)$, respectively. Assume that conditions (e) and (g) hold for a fixed $m \in \mathbb{N}$. Assume further that*

$$st_A - \lim_n \|L_n(e_0) - e_0\| = 0 \tag{11.16}$$

and

$$st_A - \lim_n \left\| L_n \left((\varphi_{\mathbf{x}})^{m+1} \right) \right\| = 0. \tag{11.17}$$

Then we get, for all X satisfying (e) and (g),

$$st_A - \lim_n \|E\left(|M_n(X) - X| \right)\| = 0. \tag{11.18}$$

Proof. Let $m \in \mathbb{N}$ be fixed. Then, it follows from Theorem 4 of [12] that, for all X satisfying (e) and (g) and for every $n \in \mathbb{N}$,

$$\|E\left(|M_n(X) - X|\right)\| \le \|E\left(|X|\right)\| \, \|L_n(e_0) - e_0\|$$
$$+ \sum_{i=1}^{m} \left(\left(\sum_{|\alpha|=i} \mu_i(\alpha) \right) \left\| L_n\left((\varphi_{\mathbf{x}})^i\right) \right\| \right)$$
$$+ \xi(m) \left\| (L_n(e_0))^{\frac{1}{m+1}} + \tau(m) \right\| \left\| L_n\left((\varphi_{\mathbf{x}})^{m+1}\right) \right\|^{\frac{m}{m+1}}$$
$$\times \left\{ \max_{\alpha:|\alpha|=m} \Omega_1\left(X_\alpha; \left\| L_n\left((\varphi_{\mathbf{x}})^{m+1}\right) \right\|^{\frac{1}{m+1}} \right)_{L_1} \right\},$$

where

$$\mu_i(\alpha) := \frac{\|E\left(|X_\alpha|\right)\|}{i!} \quad \text{for} \quad i = 1, ..., m$$

and

$$\xi(m) := \frac{1}{m!}, \quad \tau(m) := \frac{1}{m+1}.$$

As in the proof of Theorem 11.3, we deduce from (11.16) that the sequence $(\|L_n(e_0)\|)$ is A-statistically bounded, i.e., there exists a subsequence K of $\mathbb{N}$ with A-density one such that, for every $n \in K$, $\|L_n(e_0)\| \le M$ holds for some $M > 0$. Then, the above inequality gives that, for every $n \in K$,

$$\|E\left(|M_n(X) - X|\right)\| \le \|E\left(|X|\right)\| \, \|L_n(e_0) - e_0\|$$
$$+ \sum_{i=1}^{m} \left(\left(\sum_{|\alpha|=i} \mu_i(\alpha) \right) \left\| L_n\left((\varphi_{\mathbf{x}})^i\right) \right\| \right)$$
$$+ \xi(m) \left(M^{\frac{1}{m+1}} + \tau(m) \right) \left\| L_n\left((\varphi_{\mathbf{x}})^{m+1}\right) \right\|^{\frac{m}{m+1}}$$
$$\times \left\{ \max_{\alpha:|\alpha|=m} \Omega_1\left(X_\alpha; \left\| L_n\left((\varphi_{\mathbf{x}})^{m+1}\right) \right\|^{\frac{1}{m+1}} \right)_{L_1} \right\}.$$

Now applying Hölder's inequality with $\alpha = \frac{m+1}{m+1-i}$ and $\beta = \frac{m+1}{i}$, $\frac{1}{\alpha} + \frac{1}{\beta} = 1$, we may write that

$$\left\| L_n(\varphi_{\mathbf{x}}^i) \right\| \le \|L_n(e_0)\|^{\frac{m+1-i}{m+1}} \left\| L_n\left((\varphi_{\mathbf{x}})^{m+1}\right) \right\|^{\frac{i}{m+1}}, \quad i = 1, 2, ..., m.$$

Then, combining the above inequalities, we get

$$\|E\left(|M_n(X) - X|\right)\| \le \|E\left(|X|\right)\| \, \|L_n(e_0) - e_0\|$$
$$+ \sum_{i=1}^{m} \left(\left(\sum_{|\alpha|=i} \mu_i(\alpha) \right) \|L_n(e_0)\|^{\frac{m+1-i}{m+1}} \left\| L_n\left((\varphi_{\mathbf{x}})^{m+1}\right) \right\|^{\frac{i}{m+1}} \right)$$
$$+ \xi(m) \left(M^{\frac{1}{m+1}} + \tau(m) \right) \left\| L_n\left((\varphi_{\mathbf{x}})^{m+1}\right) \right\|^{\frac{m}{m+1}}$$
$$\times \left\{ \max_{\alpha:|\alpha|=m} \Omega_1\left(X_\alpha; \left\| L_n\left((\varphi_{\mathbf{x}})^{m+1}\right) \right\|^{\frac{1}{m+1}} \right)_{L_1} \right\},$$

which yields, for every $n \in K$, that

$$
\begin{aligned}
\|E\left(|M_n(X) - X|\right)\| \leq & \; A\|L_n(e_0) - e_0\| \\
& + B(m) \sum_{i=1}^{m} \left\|L_n\left((\varphi_{\mathbf{x}})^{m+1}\right)\right\|^{\frac{i}{m+1}} \\
& + C(m) \left\|L_n\left((\varphi_{\mathbf{x}})^{m+1}\right)\right\|^{\frac{m}{m+1}} \\
& \times \left\{ \max_{\boldsymbol{\alpha}:|\boldsymbol{\alpha}|=m} \Omega_1\left(X_{\boldsymbol{\alpha}}; \left\|L_n\left((\varphi_{\mathbf{x}})^{m+1}\right)\right\|^{\frac{1}{m+1}}\right)_{L_1} \right\},
\end{aligned}
$$

where

$$
A := \|E\left(|X|\right)\|, \; B(m) := \max_{i=1,..,m} \left\{ \left(\sum_{|\boldsymbol{\alpha}|=i} \mu_i(\boldsymbol{\alpha}) \right) M^{\frac{m+1-i}{m+1}} \right\}
$$

and

$$
C(m) := \xi(m)\left(M^{\frac{1}{m+1}} + \tau(m)\right).
$$

Notice that the constants A, $B(m)$ and $C(m)$ are finite positive numbers for each fixed $m \in \mathbb{N}$. Now, as in the proof of Theorem 11.3, given $\varepsilon > 0$, consider the following sets:

$$
\begin{aligned}
D(\varepsilon) &= \{n : \|E\left(|M_n(X) - X|\right)\| \geq \varepsilon\}, \\
D_0(\varepsilon) &= \left\{ n : \|L_n(e_0) - e_0\| \geq \frac{\varepsilon}{(m+3)A} \right\}, \\
D_i(\varepsilon) &= \left\{ n : \left\|L_n\left((\varphi_{\mathbf{x}})^{m+1}\right)\right\|^{\frac{i}{m+1}} \geq \frac{\varepsilon}{(m+3)B(m)} \right\}, \quad i = 1, 2, ..., m, \\
D_{m+1}(\varepsilon) &= \left\{ n : \left\|L_n\left((\varphi_{\mathbf{x}})^{m+1}\right)\right\|^{\frac{m}{m+1}} \geq \sqrt{\frac{2\varepsilon}{(m+3)C(m)}} \right\}, \\
D_{m+2}(\varepsilon) &= \left\{ n : \max_{\boldsymbol{\alpha}:|\boldsymbol{\alpha}|=m} \Omega_1\left(X_{\boldsymbol{\alpha}}; \left\|L_n\left((\varphi_{\mathbf{x}})^{m+1}\right)\right\|^{\frac{1}{m+1}}\right)_{L_1} \geq \sqrt{\frac{2\varepsilon}{(m+3)C(m)}} \right\}.
\end{aligned}
$$

Then, in order to obtain (11.18) we perform the same lines as in proof of Theorem 11.3. ∎

11.2 Conclusions

In this section, we present some consequences of Theorems 11.3 and 11.4. We also give a sequence of positive linear operators which satisfies the statistical approximation results obtained in this chapter but not their classical cases.

If we replace the matrix $A = [a_{jn}]$ by the identity matrix in Theorems 11.3 and 11.4, then we immediately get the next results introduced by Anastassiou [12].

Corollary 11.5 (see [12]). *Let (L_n), $X(x, \omega)$, (M_n) and E be the same as in $(a) - (d)$, respectively. Assume that conditions (e) and (f) hold for a fixed $m \in \mathbb{N}$ and a fixed $q \in (1, \infty)$. Assume further that the sequences $(L_n(e_0))$ and $\left(L_n \left((\varphi_\mathbf{x})^{q(m+1)} \right) \right)$ are uniformly convergent to e_0 and 0 on a compact and convex subset $\mathbf{Q}$ of $\mathbb{R}^k$ $(k > 1)$, respectively. Then, for all X satisfying (e) and (f), $(E\left(|M_n(X) - X|^q \right))$ is uniformly convergent to 0 on $\mathbf{Q}$.*

Corollary 11.6 (see [12]). *Let (L_n), $X(x, \omega)$, (M_n) and E be the same as in $(a)-(d)$, respectively. Assume that conditions (e) and (g) hold for a fixed $m \in \mathbb{N}$. Assume further that the sequences $(L_n(e_0))$ and $\left(L_n \left((\varphi_\mathbf{x})^{m+1} \right) \right)$ are uniformly convergent to e_0 and 0 on a compact and convex subset $\mathbf{Q}$ of $\mathbb{R}^k$ $(k > 1)$, respectively. Then, for all X satisfying (e) and (g), $\{E\left(|M_n(X) - X| \right)\}$ is uniformly convergent to 0 on $\mathbf{Q}$.*

Taking $m = 0$ in Theorem 11.4, we obtain the next statistical Korovkin-type result for multivariate stochastic processes.

Corollary 11.7. *Let $A = [a_{jn}]$ be a non-negative regular summability matrix, and let (L_n), $X(x, \omega)$, (M_n) and E be the same as in $(a) - (d)$, respectively. Assume that conditions (e) and (g) hold for $m = 0$ and $\mathbf{Q} := [a_1, b_1] \times \dots \times [a_k, b_k] \subset \mathbb{R}^k$ $(k > 1)$. Assume further that, for each $i = 0, 1, \dots, 2k$*

$$st_A - \lim_n \| L_n(e_k) - e_k \| = 0, \qquad (11.19)$$

where

$$e_0(\mathbf{x}) = 1, \ e_j(\mathbf{x}) = x_j \ \text{ and } e_{k+j}(\mathbf{x}) = x_j^2 \ (j = 1, 2, \dots, k). \qquad (11.20)$$

Then we get, for all X satisfying (e) and (g),

$$st_A - \lim_n \| E\left(|M_n(X) - X| \right) \| = 0.$$

Proof. By (11.19) there exists a subset K of $\mathbb{N}$ with A-density one such that $\| L_n(e_0) \| \leq M$ holds for every $n \in K$ and for some $M > 0$. Let $\mathbf{x} \in \mathbf{Q}$.

Then, using the Cauchy-Schwarz inequality, we see, for every $n \in K$, that

$$L_n\left(\varphi_{\mathbf{x}}; \mathbf{x}\right) = L_n\left(\|\mathbf{y} - \mathbf{x}\|_{\ell_1}; \mathbf{x}\right)$$

$$= L_n\left(\sum_{i=1}^{m} |y_i - x_i|; \mathbf{x}\right)$$

$$= \sum_{i=1}^{m} L_n\left(|y_i - x_i|; \mathbf{x}\right)$$

$$\leq \sqrt{M} \sum_{i=1}^{m} \left\{L_n\left((y_i - x_i)^2; \mathbf{x}\right)\right\}^{\frac{1}{2}}.$$

Since, for each $i = 1, 2, ..., m$,

$$L_n\left((y_i - x_i)^2; \mathbf{x}\right) = L_n(e_{k+i}; \mathbf{x}) - 2x_i L_n(e_i; \mathbf{x}) + x_i^2 L_n(e_0; \mathbf{x})$$

$$= (L_n(e_{k+i}; \mathbf{x}) - e_{k+i}(\mathbf{x})) - 2x_i (L_n(e_i; \mathbf{x}) - e_i(\mathbf{x}))$$

$$+ x_i^2 (L_n(e_0; \mathbf{x}) - e_0(\mathbf{x}))$$

$$\leq |L_n(e_{k+i}; \mathbf{x}) - e_{k+i}(\mathbf{x})| + 2C |L_n(e_i; \mathbf{x}) - e_i(\mathbf{x})|$$

$$+ C^2 |L_n(e_0; \mathbf{x}) - e_0(\mathbf{x})|,$$

where $C := \max\{|x_1|, |x_2|, ..., |x_m|\}$, we immediately observe that, for every $n \in K$,

$$L_n\left(\varphi_{\mathbf{x}}; \mathbf{x}\right) \leq \sqrt{M} \sum_{i=1}^{m} \{|L_n(e_{k+i}; \mathbf{x}) - e_{k+i}(\mathbf{x})| + 2C |L_n(e_i; \mathbf{x}) - e_i(\mathbf{x})|$$

$$+ C^2 |L_n(e_0; \mathbf{x}) - e_0(\mathbf{x})|\}^{\frac{1}{2}},$$

which implies that

$$\|L_n\left(\varphi_{\mathbf{x}}\right)\| \leq \sqrt{M} \sum_{i=1}^{m} \{\|L_n(e_{k+i}) - e_{k+i}\| + 2C \|L_n(e_i) - e_i\|$$

$$+ C^2 \|L_n(e_0) - e_0\|\}^{\frac{1}{2}}$$

$$\leq \sqrt{M} \sum_{i=1}^{m} \left\{\|L_n(e_{k+i}) - e_{k+i}\|^{\frac{1}{2}} + \sqrt{2C} \|L_n(e_i) - e_i\|^{\frac{1}{2}}\right.$$

$$\left. + C \|L_n(e_0) - e_0\|^{\frac{1}{2}}\right\}.$$

By the last inequality and (11.19), it is easy to check that

$$st_A - \lim_n \|L_n\left(\varphi_{\mathbf{x}}\right)\| = 0.$$

Hence, all hypotheses of Theorem 11.4 are fulfilled for $m = 0$. So, the proof is done. ∎

Of course, if we take $A = I$, the identity matrix, then the following result immediately follows from Corollary 11.7.

Corollary 11.8. *Let* (L_n), $X(x, \omega)$, (M_n) *and* E *be the same as in* (a) − (d), *respectively. Assume that conditions* (e) *and* (g) *hold for* $m = 0$ *and* $\mathbf{Q} := [a_1, b_1] \times ... \times [a_k, b_k] \subset \mathbb{R}^k$ $(k > 1)$. *Assume further that, for each* $i = 0, 1, ..., 2k$, *the sequence* $(L_n(e_k))$ *is uniformly convergent to* e_k *on* $\mathbf{Q}$. *Then, for all* X *satisfying* (e) *and* (g), *the sequence* $(E(|M_n(X) - X|))$ *is uniformly convergent to* 0 *on* $\mathbf{Q}$.

Finally, we give an example as follows. Let $\mathbf{Q} := [0, 1] \times [0, 1]$. Define a sequence (u_n) by

$$u_n := \begin{cases} 0, & \text{if } n = k^2, \ k \in \mathbb{N} \\ 1, & \text{otherwise.} \end{cases}$$

Then, we know $st_{C_1} - \lim_n u_n = st - \lim_n u_n = 1$ although (u_n) is non-convergent in the usual sense. Now, define the next positive linear operators:

$$L_n(f; \mathbf{x}) = \sum_{k=0}^{n} \sum_{l=0}^{n} f\left(\frac{k}{n}, \frac{l}{n}\right) \binom{n}{k} \binom{n}{l} u_n^{k+l} x_1^k x_2^l (1 - u_n x_1)^{n-k} (1 - u_n x_2)^{n-l},$$

(11.21)

where $\mathbf{x} = (x_1, x_2) \in \mathbf{Q} = [0, 1] \times [0, 1]$, $f \in C(\mathbf{Q})$ and $n \in \mathbb{N}$. Furthermore, define $M_n(X)$ as follows:

$$\begin{aligned} M_n(X)(\mathbf{x}, \omega) :&= L_n(X(\cdot, \omega); \mathbf{x}) \\ &= \sum_{k=0}^{n} \sum_{l=0}^{n} X\left(\frac{k}{n}, \frac{l}{n}, \omega\right) \binom{n}{k} \binom{n}{l} \\ &\quad \cdot u_n^{k+l} x_1^k x_2^l (1 - u_n x_1)^{n-k} (1 - u_n x_2)^{n-l}, \end{aligned}$$

(11.22)

where $\mathbf{x} = (x_1, x_2) \in \mathbf{Q}$, $\omega \in \Omega$, $n \in \mathbb{N}$, and X is a stochastic process in two variables satisfying (e) and (g) for $m = 0$. In this case, using the bivariate Bernstein polynomials given by

$$B_n(f; \mathbf{x}) := \sum_{k=0}^{n} \sum_{l=0}^{n} f\left(\frac{k}{n}, \frac{l}{n}\right) \binom{n}{k} \binom{n}{l} x_1^k x_2^l (1 - x_1)^{n-k} (1 - x_2)^{n-l},$$

we obtain from (11.21) that

$$L_n(f; \mathbf{x}) = B_n(f; u_n \mathbf{x}).$$

(11.23)

Then, by (11.23), we see that

$$\begin{aligned} L_n(e_0; \mathbf{x}) &= 1, \\ L_n(e_1; \mathbf{x}) &= u_n e_1(\mathbf{x}), \\ L_n(e_2; \mathbf{x}) &= u_n e_2(\mathbf{x}), \\ L_n(e_3; \mathbf{x}) &= u_n^2 e_3(\mathbf{x}) + \frac{u_n x_1(1 - u_n x_1)}{n}, \\ L_n(e_4; \mathbf{x}) &= u_n^2 e_4(\mathbf{x}) + \frac{u_n x_2(1 - u_n x_2)}{n} \end{aligned}$$

where the functions e_i $(i = 0, 1, 2, 3, 4)$ are given by (11.20). So, by the definition of (u_n), we easily get that

$$st - \lim_n \|L_n(e_k) - e_k\| = 0, \quad (i = 0, 1, 2, 3, 4).$$

Hence, by (11.22) and Corollary 11.7, if X satisfies (e) and (g) for $m = 0$, then

$$st - \lim_n \|E\left(|M_n(X) - X|\right)\| = 0.$$

However, since the sequence (u_n) is non-convergent, Corollary 11.8 which is the classical version of Corollary 11.7 fails. Therefore, this application demonstrates that the statistical approximation results for multivariate stochastic processes are more applicable than their classical cases introduced in [12].

12

Fractional Korovkin-Type Approximation Theory Based on Statistical Convergence

In this chapter, we get some statistical Korovkin-type approximation theorems including fractional derivatives of functions. Furthermore, we demonstrate that these results are more applicable than the classical ones. This chapter relies on [21].

12.1 Fractional Derivatives and Positive Linear Operators

In this section we first recall the Caputo fractional derivatives. Let r be a positive real number and $m = \lceil r \rceil$, where $\lceil \cdot \rceil$ is the ceiling of the number. As usual, by $AC([a, b])$ we denote the space of all real-valued absolutely continuous functions on $[a, b]$. Also, consider the space

$$AC^m([a, b]) := \left\{ f : [a, b] \to \mathbb{R} : f^{(m-1)} \in AC([a, b]) \right\}.$$

Then, the left Caputo fractional derivative?? of a function f belonging to $AC^m[a, b]$ is defined by

$$D_{*a}^r f(x) := \frac{1}{\Gamma(m - r)} \int_a^x (x - t)^{m-r-1} f^{(m)}(t) dt \quad \text{for} \quad x \in [a, b], \quad (12.1)$$

G.A. Anastassiou and O. Duman: Towards Intelligent Modeling, ISRL 14, pp. 157–167.
springerlink.com © Springer-Verlag Berlin Heidelberg 2011

where Γ is the usual Gamma function. Also, the right Caputo fractional derivative of a function f belonging to $AC^m([a,b])$ is defined by

$$D_{b-}^r f(x) := \frac{(-1)^m}{\Gamma(m-r)} \int_x^b (\zeta - x)^{m-r-1} f^{(m)}(\zeta) d\zeta \quad \text{for} \quad x \in [a,b]. \quad (12.2)$$

In (12.1) and (12.2), we set $D_{*a}^0 f = f$ and $D_{b-}^0 f = f$ on $[a,b]$. Throughout this section, we assume that

$$D_{*a}^r f(y) = 0 \quad \text{for every } y < a$$

and

$$D_{b-}^r f(y) = 0 \quad \text{for every } y > b.$$

Then we know the following facts (see, e.g., [13, 14, 100, 103]):

(a) If $r > 0$, $r \notin \mathbb{N}$, $m = \lceil r \rceil$, $f \in C^{m-1}([a,b])$ and $f^{(m)} \in L_\infty([a,b])$, then we get $D_{*a}^r f(a) = 0$ and $D_{b-}^r f(b) = 0$.

(b) Let $y \in [a,b]$ be fixed. For $r > 0$, $m = \lceil r \rceil$, $f \in C^{m-1}([a,b])$ with $f^{(m)} \in L_\infty[a,b]$, define the following Caputo fractional derivatives:

$$U_f(x,y) := D_{*x}^r f(y) = \frac{1}{\Gamma(m-r)} \int_x^y (y-t)^{m-r-1} f^{(m)}(t) dt \quad \text{for} \quad y \in [x,b]$$

$$(12.3)$$

and

$$V_f(x,y) := D_{x-}^r f(y) = \frac{(-1)^m}{\Gamma(m-r)} \int_y^x (\zeta - y)^{m-r-1} f^{(m)}(\zeta) d\zeta \quad \text{for} \quad y \in [a,x].$$

$$(12.4)$$

Then, by [13], for each fixed $x \in [a,b]$, $U_f(x,.)$ is continuous on the interval $[x,b]$, and also $V_f(x,.)$ is continuous on $[a,x]$. In addition, if $f \in C^m([a,b])$, then, $U_f(\cdot,\cdot)$ and $V_f(\cdot,\cdot)$ are continuous on the set $[a,b] \times [a,b]$.

(c) Let $\omega(f,\delta)$, $\delta > 0$, denote the usual modulus of continuity of a function f on $[a,b]$. If $g \in C([a,b] \times [a,b])$, then, for any $\delta > 0$, both the functions $s(x) := \omega(g(.,x),\delta)_{[a,x]}$ and $t(x) := \omega(g(.,x),\delta)_{[x,b]}$ are continuous at the point $x \in [a,b]$.

(d) If $f \in C^{m-1}([a,b])$ with $f^{(m)} \in L_\infty[a,b]$, then we obtain from [13] that

$$\sup_{x \in [a,b]} \omega(U_f(x,\cdot),\delta)_{[x,b]} < \infty \quad (12.5)$$

and

$$\sup_{x \in [a,b]} \omega \left(V_f \left(x, \cdot \right), \delta \right)_{[a,x]} < \infty. \tag{12.6}$$

(e) Now let $\Psi(y) := \Psi_x(y) = y - x$ and $e_0(y) := 1$ on the interval $[a,b]$. Following the paper by Anastassiou (see [13]) if $L_n : C([a,b]) \to C([a,b])$ is sequence of positive linear operators and if $r > 0$, $r \notin \mathbb{N}$, $m = \lceil r \rceil$, $f \in AC^m([a,b])$ with $f^{(m)} \in L_\infty([a,b])$, then we get that ($\|.\|$ is the sup-norm)

$$\|L_n(f) - f\| \leq \|f\| \, \|L_n(e_0) - e_0\| + \sum_{k=1}^{m-1} \frac{\|f^{(k)}\|}{k!} \left\| L_n \left(|\Psi|^k \right) \right\|$$

$$+ \left(\frac{r+2}{\Gamma(r+2)} + \frac{1}{\Gamma(r+1)} \|L_n(e_0) - e_0\|^{\frac{1}{r+1}} \right)$$

$$\times \left\| L_n \left(|\Psi|^{r+1} \right) \right\|^{\frac{r}{r+1}} \left\{ \sup_{x \in [a,b]} \omega \left(U_f(x, \cdot), \left\| L_n \left(|\Psi|^{r+1} \right) \right\|^{\frac{1}{r+1}} \right)_{[x,b]} \right.$$

$$\left. + \sup_{x \in [a,b]} \omega \left(V_f(x, \cdot), \left\| L_n \left(|\Psi|^{r+1} \right) \right\|^{\frac{1}{r+1}} \right)_{[a,x]} \right\}.$$

Then letting

$$\delta_{n,r} := \left\| L_n \left(|\Psi|^{r+1} \right) \right\|^{\frac{1}{r+1}}, \tag{12.7}$$

and also considering (12.5), (12.6) we can write that

$$\|L_n(f) - f\| \leq K_{m,r} \left\{ \|L_n(e_0) - e_0\| + \sum_{k=1}^{m-1} \left\| L_n \left(|\Psi|^k \right) \right\| \right.$$

$$+ \delta_{n,r}^r \left(\sup_{x \in [a,b]} \omega \left(U_f(x, \cdot), \delta_{n,r} \right)_{[x,b]} \right)$$

$$+ \delta_{n,r}^r \left(\sup_{x \in [a,b]} \omega \left(V_f(x, \cdot), \delta_{n,r} \right)_{[a,x]} \right)$$

$$+ \delta_{n,r}^r \|L_n(e_0) - e_0\|^{\frac{1}{r+1}} \left(\sup_{x \in [a,b]} \omega \left(U_f(x, \cdot), \delta_{n,r} \right)_{[x,b]} \right)$$

$$\left. + \delta_{n,r}^r \|L_n(e_0) - e_0\|^{\frac{1}{r+1}} \left(\sup_{x \in [a,b]} \omega \left(V_f(x, \cdot), \delta_{n,r} \right)_{[a,x]} \right) \right\}, \tag{12.8}$$

where

$$K_{r,m} := \max \left\{ \frac{1}{\Gamma(r+1)}, \frac{r+2}{\Gamma(r+2)}, \|f\|, \|f'\|, \frac{\|f''\|}{2!}, \frac{\|f'''\|}{3!}, ..., \frac{\left\| f^{(m-1)} \right\|}{(m-1)!} \right\}. \tag{12.9}$$

Notice that the sum in the right hand-side of (12.8) collapses when $r \in (0, 1)$.

Thus, the following theorem is a fractional Korovkin-type approximation result for a sequence of positive linear operators.

Theorem A (see [13]). *Let $L_n : C([a, b]) \to C([a, b])$ be a sequence of positive linear operators, and let $r > 0$, $r \notin \mathbb{N}$, $m = \lceil r \rceil$. If the sequence $(\delta_{n,r})$ given by (12.7) is convergent to zero as n tends to infinity and $(L_n(e_0))$ converges uniformly to e_0 on $[a, b]$, then, for every $f \in AC^m([a, b])$ with $f^{(m)} \in L_\infty([a, b])$, the sequence $(L_n(f))$ converges uniformly to f on the interval $[a, b]$. Furthermore, this uniform convergence is still valid on $[a, b]$ when $f \in C^m([a, b])$.*

12.2 Fractional Korovkin Results Based on Statistical Convergence

In this section, we mainly get the statistical version of Theorem A. At first we need the next lemma.

Lemma 12.1. *Let $A = [a_{jn}]$ be a non-negative regular summability matrix, and let $r > 0$, $r \notin \mathbb{N}$, $m = \lceil r \rceil$. Assume that $L_n : C([a, b]) \to C([a, b])$ is a sequence of positive linear operators. If*

$$st_A - \lim_n \|L_n(e_0) - e_0\| = 0 \tag{12.10}$$

and

$$st_A - \lim_n \delta_{n,r} = 0, \tag{12.11}$$

where $\delta_{n,r}$ is the same as in (12.7), then we get, for every $k = 1, 2, ..., m-1$,

$$st_A - \lim_n \left\| L_n \left(|\Psi|^k \right) \right\| = 0.$$

Proof. Let $k \in \{1, 2, ..., m - 1\}$ be fixed. Applying Hölder's inequality for positive linear operators with $p = \dfrac{r + 1}{k}$, $q = \dfrac{r + 1}{r + 1 - k}$ $\left(\dfrac{1}{p} + \dfrac{1}{q} = 1 \right)$, we see that

$$\left\| L_n \left(|\Psi|^k \right) \right\| \leq \left\| L_n \left(|\Psi|^{r+1} \right) \right\|^{\frac{k}{r+1}} \|L_n(e_0)\|^{\frac{r+1-k}{r+1}} ,$$

which implies

$$\left\| L_n \left(|\Psi|^k \right) \right\| \leq \left\| L_n \left(|\Psi|^{r+1} \right) \right\|^{\frac{k}{r+1}} \left\{ \|L_n(e_0) - e_0\|^{\frac{r+1-k}{r+1}} + 1 \right\}.$$

Then, for each $k = 1, 2, ..., m - 1$, we obtain the following inequality

$$\left\| L_n \left(|\Psi|^k \right) \right\| \leq \delta_{n,r}^k \left\| L_n \left(e_0 \right) - e_0 \right\|^{\frac{r+1-k}{r+1}} + \delta_{n,r}^k. \tag{12.12}$$

Now, for a given $\varepsilon > 0$, consider the following sets:

$$A := \left\{ n \in \mathbb{N} : \left\| L_n \left(|\Psi|^k \right) \right\| \geq \varepsilon \right\},$$

$$A_1 := \left\{ n \in \mathbb{N} : \delta_{n,r}^k \left\| L_n \left(e_0 \right) - e_0 \right\|^{\frac{r+1-k}{r+1}} \geq \frac{\varepsilon}{2} \right\}$$

$$A_2 := \left\{ n \in \mathbb{N} : \delta_{n,r} \geq \left(\frac{\varepsilon}{2} \right)^{\frac{1}{k}} \right\}.$$

Then, it follows from (12.12) that $A \subseteq A_1 \cup A_2$. Also, defining

$$A_1' := \left\{ n \in \mathbb{N} : \delta_{n,r} \geq \left(\frac{\varepsilon}{2} \right)^{\frac{1}{2k}} \right\},$$

$$A_1'' := \left\{ n \in \mathbb{N} : \left\| L_n \left(e_0 \right) - e_0 \right\| \geq \left(\frac{\varepsilon}{2} \right)^{\frac{r+1}{2(r+1-k)}} \right\},$$

we derive that $A_1 \subseteq A_1' \cup A_2''$, which gives

$$A \subseteq A_1' \cup A_1'' \cup A_2.$$

Hence, for every $j \in \mathbb{N}$, we observe that

$$\sum_{n \in A} a_{jn} \leq \sum_{n \in A_1'} a_{jn} + \sum_{n \in A_1''} a_{jn} + \sum_{n \in A_2} a_{jn}.$$

Taking limit as $j \to \infty$ in the last inequality and also using the hypotheses (12.10) and (12.11) we immediately obtain that

$$\lim_j \sum_{n \in A} a_{jn} = 0.$$

Hence, we deduce that, for each $k = 1, 2, ..., m - 1$,

$$st_A - \lim_n \left\| L_n \left(|\Psi|^k \right) \right\| = 0,$$

which completes the proof.   ∎

Now we are ready to give the following fractional approximation result based on statistical convergence.

Theorem 12.2. *Let $A = [a_{jn}]$ be a non-negative regular summability matrix, and let $r > 0$, $r \notin \mathbb{N}$, $m = \lceil r \rceil$. Assume that $L_n : C([a, b]) \to C([a, b])$ is a sequence of positive linear operators. If (12.10) and (12.11) hold, then, for every $f \in AC^m([a, b])$ with $f^{(m)} \in L_\infty([a, b])$, we get*

$$st_A - \lim_n \|L_n(f) - f\| = 0. \tag{12.13}$$

Proof. Let $f \in AC^m([a,b])$ with $f^{(m)} \in L_\infty([a,b])$. Then, using (12.5), (12.6) and (12.8), we have

$$\|L_n(f) - f\| \leq M_{m,r} \Big\{ \|L_n(e_0) - e_0\| + 2\delta_{n,r}^r$$
$$+ 2\delta_{n,r}^r \|L_n(e_0) - e_0\|^{\frac{1}{r+1}} + \sum_{k=1}^{m-1} \left\| L_n\left(|\Psi|^k\right) \right\| \Big\}, \qquad (12.14)$$

where

$$M_{m,r} := \max \left\{ K_{m,r}, \; \sup_{x \in [a,b]} \omega\left(U_f(x,\cdot), \delta_{n,r}\right)_{[x,b]}, \; \sup_{x \in [a,b]} \omega\left(V_f(x,\cdot), \delta_{n,r}\right)_{[a,x]} \right\}$$

and $K_{m,r}$ is given by (12.9). Now, for a given $\varepsilon > 0$, consider the following sets:

$$B := \{ n \in \mathbb{N} : \|L_n(f) - f\| \geq \varepsilon \},$$

$$B_k := \left\{ n \in \mathbb{N} : \left\| L_n\left(|\Psi|^k\right) \right\| \geq \frac{\varepsilon}{(m+2)M_{m,r}} \right\}, \quad k = 1, 2, ..., m-1.$$

$$B_m := \left\{ n \in \mathbb{N} : \|L_n(e_0) - e_0\| \geq \frac{\varepsilon}{(m+2)M_{m,r}} \right\}$$

$$B_{m+1} := \left\{ n \in \mathbb{N} : \delta_{n,r} \geq \left(\frac{\varepsilon}{2(m+2)M_{m,r}} \right)^{\frac{1}{r}} \right\},$$

$$B_{m+2} := \left\{ n \in \mathbb{N} : \delta_{n,r}^r \|L_n(e_0) - e_0\|^{\frac{1}{r+1}} \geq \frac{\varepsilon}{2(m+2)M_{m,r}} \right\}.$$

Then, it follows from (12.14) that $B \subseteq \bigcup_{i=1}^{m+2} B_i$. Also defining

$$B_{m+3} := \left\{ n \in \mathbb{N} : \|L_n(e_0) - e_0\| \geq \left(\frac{\varepsilon}{2(m+2)M_{m,r}} \right)^{\frac{r+1}{2}} \right\},$$

$$B_{m+4} := \left\{ n \in \mathbb{N} : \delta_{n,r} \geq \left(\frac{\varepsilon}{2(m+2)M_{m,r}} \right)^{\frac{1}{2r}} \right\}$$

we observe that

$$B_{m+2} \subseteq B_{m+3} \cup B_{m+4},$$

which yields that

$$B \subseteq \bigcup_{i=1 \; (i \neq m+2)}^{m+4} B_i.$$

Hence, for every $j \in \mathbb{N}$, we obtain that

$$\sum_{n \in B} a_{jn} \leq \sum_{i=1 \; (i \neq m+2)}^{m+4} \sum_{n \in B_i} a_{jn}. \qquad (12.15)$$

Letting $j \to \infty$ in the both sides of (12.15) and also using (12.10), (12.11), and also considering Lemma 12.1 we deduce that

$$\lim_j \sum_{n \in B} a_{jn} = 0,$$

which implies (12.13).   ∎

If we take the space $C^m([a,b])$ instead of $AC^m([a,b])$, then we can obtain a slight modification of Theorem 12.2. To see this we need the following lemma.

Lemma 12.3. *Let $A = [a_{jn}]$ be a non-negative regular summability matrix, and let $r > 0$, $r \notin \mathbb{N}$, $m = \lceil r \rceil$. Assume that $L_n : C([a,b]) \to C([a,b])$ is a sequence of positive linear operators. If (12.11) holds, then, for every $f \in C^m([a,b])$, we get:*

(i) $st_A - \lim_n \left(\sup_{x \in [a,b]} \omega \left(U_f(x,\cdot), \delta_{n,r} \right)_{[x,b]} \right) = 0,$

(ii) $st_A - \lim_n \left(\sup_{x \in [a,b]} \omega \left(V_f(x,\cdot), \delta_{n,r} \right)_{[a,x]} \right) = 0,$

where $\delta_{n,r}$ is the same as in (12.7); $U_f(\cdot,\cdot)$ and $V_f(\cdot,\cdot)$ are given respectively by (12.3) and (12.4).

Proof. From **(b)** if $f \in C^m([a,b])$, then both $U_f(\cdot,\cdot)$ and $V_f(\cdot,\cdot)$ belong to $C([a,b] \times [a,b])$. Then, by **(c)**, the functions $\omega \left(U_f(x,\cdot), \delta_{n,r} \right)_{[x,b]}$ and $\omega \left(V_f(x,\cdot), \delta_{n,r} \right)_{[a,x]}$ are continuous at the point $x \in [a,b]$. Thus, there exist the points $x_0, x_1 \in [a,b]$ such that

$$\sup_{x \in [a,b]} \omega \left(U_f(x,\cdot), \delta_{n,r} \right)_{[x,b]} = \omega \left(U_f(x_0,\cdot), \delta_{n,r} \right)_{[x_0,b]} =: g\left(\delta_{n,r} \right)$$

and

$$\sup_{x \in [a,b]} \omega \left(V_f(x,\cdot), \delta_{n,r} \right)_{[a,x]} = \omega \left(V_f(x_1,\cdot), \delta_{n,r} \right)_{[a,x_1]} =: h\left(\delta_{n,r} \right).$$

Since $U_f(x_0,\cdot)$ and $V_f(x_1,\cdot)$ are continuous on $[a,b]$, the functions g and h are right continuous at the origin. By (12.11), we obtain, for any $\delta > 0$, that

$$\lim_j \sum_{n:\delta_{n,r} \geq \delta} a_{jn} = 0. \tag{12.16}$$

Now, by the right continuity of g and h at zero, for a given $\varepsilon > 0$, there exist $\delta_1, \delta_2 > 0$ such that $g(\delta_{n,r}) < \varepsilon$ whenever $\delta_{n,r} < \delta_1$ and that $h(\delta_{n,r}) < \varepsilon$

whenever $\delta_{n,r} < \delta_2$. Then, we can write that $g(\delta_{n,r}) \geq \varepsilon$ gives $\delta_{n,r} \geq \delta_1$, and also that $h(\delta_{n,r}) \geq \varepsilon$ yields $\delta_{n,r} \geq \delta_2$. Hence, we observe that

$$\{n \in \mathbb{N} : g(\delta_{n,r}) \geq \varepsilon\} \subseteq \{n \in \mathbb{N} : \delta_{n,r} \geq \delta_1\} \tag{12.17}$$

and

$$\{n \in \mathbb{N} : h(\delta_{n,r}) \geq \varepsilon\} \subseteq \{n \in \mathbb{N} : \delta_{n,r} \geq \delta_2\} \tag{12.18}$$

So, it follows from (12.17) and (12.18) that, for each $j \in \mathbb{N}$,

$$\sum_{n:g(\delta_{n,r}) \geq \varepsilon} a_{jn} \leq \sum_{n:\delta_{n,r} \geq \delta_1} a_{jn} \tag{12.19}$$

and

$$\sum_{n:h(\delta_{n,r}) \geq \varepsilon} a_{jn} \leq \sum_{n:\delta_{n,r} \geq \delta_2} a_{jn} \tag{12.20}$$

Then, letting $j \to \infty$ on the both sides of the inequalities (12.19), (12.20); and also using (12.16) we immediately see, for every $\varepsilon > 0$,

$$\lim_j \sum_{n:g(\delta_{n,r}) \geq \varepsilon} a_{jn} = \lim_j \sum_{n:h(\delta_{n,r}) \geq \varepsilon} a_{jn} = 0,$$

which implies that

$$st_A - \lim_n \left(\sup_{x \in [a,b]} \omega \left(U_f(x,\cdot), \delta_{n,r} \right)_{[x,b]} \right) = 0$$

and

$$st_A - \lim_n \left(\sup_{x \in [a,b]} \omega \left(V_f(x,\cdot), \delta_{n,r} \right)_{[a,x]} \right) = 0.$$

Therefore, the proof of Lemma is finished. ∎

Then, we obtain the next result.

Theorem 12.4. *Let $A = [a_{jn}]$ be a non-negative regular summability matrix, and let $r > 0$, $r \notin \mathbb{N}$, $m = \lceil r \rceil$. Assume that $L_n : C([a,b]) \to C([a,b])$ is a sequence of positive linear operators. If (12.10) and (12.11) hold, then, for every $f \in C^m([a,b])$, we get (12.13).*

Proof. By (12.8), we see that

$$\begin{aligned}
\|L_n(f) - f\| \leq K_{m,r} \Big\{ &\|L_n(e_0) - e_0\| + \sum_{k=1}^{m-1} \left\|L_n\left(|\Psi|^k\right)\right\| \\
&+ \delta_{n,r}^r g(\delta_{n,r}) + \delta_{n,r}^r h(\delta_{n,r}) \\
&+ \delta_{n,r}^r g(\delta_{n,r}) \|L_n(e_0) - e_0\|^{\frac{1}{r+1}} \\
&+ \delta_{n,r}^r h(\delta_{n,r}) \|L_n(e_0) - e_0\|^{\frac{1}{r+1}} \Big\},
\end{aligned} \tag{12.21}$$

where $g(\delta_{n,r})$ and $h(\delta_{n,r})$ are the same as in the proof of Lemma 12.3. Now, for a given $\varepsilon > 0$, define the following sets

$$C := \{n \in \mathbb{N} : \|L_n(f) - f\| \geq \varepsilon\},$$

$$C_k := \left\{n \in \mathbb{N} : \left\|L_n\left(|\Psi|^k\right)\right\| \geq \frac{\varepsilon}{(m+4)K_{m,r}}\right\}, \quad k = 1, 2, ..., m-1.$$

$$C_m := \left\{n \in \mathbb{N} : \|L_n(e_0) - e_0\| \geq \frac{\varepsilon}{(m+4)K_{m,r}}\right\}$$

$$C_{m+1} := \left\{n \in \mathbb{N} : \delta_{n,r}^r g(\delta_{n,r}) \geq \frac{\varepsilon}{(m+4)K_{m,r}}\right\},$$

$$C_{m+2} := \left\{n \in \mathbb{N} : \delta_{n,r}^r h(\delta_{n,r}) \geq \frac{\varepsilon}{(m+4)K_{m,r}}\right\},$$

$$C_{m+3} := \left\{n \in \mathbb{N} : \delta_{n,r}^r g(\delta_{n,r}) \|L_n(e_0) - e_0\|^{\frac{1}{r+1}} \geq \frac{\varepsilon}{(m+4)K_{m,r}}\right\}$$

$$C_{m+4} := \left\{n \in \mathbb{N} : \delta_{n,r}^r h(\delta_{n,r}) \|L_n(e_0) - e_0\|^{\frac{1}{r+1}} \geq \frac{\varepsilon}{(m+4)K_{m,r}}\right\}.$$

Then, by (12.21), we get

$$C \subseteq \bigcup_{i=1}^{m+4} C_i.$$

So, for every $j \in \mathbb{N}$, we have

$$\sum_{n \in C} a_{jn} \leq \sum_{i=1}^{m+4} \left(\sum_{n \in C_i} a_{jn}\right). \tag{12.22}$$

On the other hand, by (12.10), (12.11) and Lemmas 12.1, 12.3, we observe that

$$st_A - \lim_n \left\|L_n\left(|\Psi|^k\right)\right\| = 0, \quad (k = 1, .., m-1),$$

$$st_A - \lim_n \delta_{n,r}^r g(\delta_{n,r}) = 0,$$

$$st_A - \lim_n \delta_{n,r}^r h(\delta_{n,r}) = 0,$$

$$st_A - \lim_n \delta_{n,r}^r g(\delta_{n,r}) \|L_n(e_0) - e_0\|^{\frac{1}{r+1}} = 0,$$

$$st_A - \lim_n \delta_{n,r}^r h(\delta_{n,r}) \|L_n(e_0) - e_0\|^{\frac{1}{r+1}} = 0.$$

Hence, we deduce that, for every $i = 1, 2, ..., m+4$,

$$\lim_j \sum_{n \in C_i} a_{jn} = 0. \tag{12.23}$$

Now, letting $j \to \infty$ in the both sides of (12.22) and using (12.23) we see that

$$\lim_j \sum_{n \in C} a_{jn} = 0.$$

The last equality gives that

$$st_A - \lim_n \|L_n(f) - f\| = 0,$$

which finishes the proof.   ∎

12.3 Conclusions

In this section we present a sequence of positive linear operators which satisfies all conditions of Theorem 12.2 but not Theorem A.

Now take $A = C_1 = [c_{jn}]$, the Cesáro matrix, and define the sequences (u_n) and (v_n) by

$$u_n := \begin{cases} \sqrt{n}, & \text{if } n = m^2 \ (m \in \mathbb{N}), \\ 0, & \text{otherwise.} \end{cases}$$

and

$$v_n := \begin{cases} 1/2, & \text{if } n = m^2 \ (m \in \mathbb{N}), \\ 1, & \text{otherwise.} \end{cases}$$

Then we easily see that

$$st - \lim_n u_n = 0 \quad \text{and} \quad st - \lim_n v_n = 1.$$

Let $r = \dfrac{1}{2}$. Then we have $m = \left\lceil \dfrac{1}{2} \right\rceil = 1$. Now define the following Bernstein-like positive linear operators:

$$L_n(f; x) := (1 + u_n) \sum_{k=0}^{n} f\left(\frac{k}{n}\right) \binom{n}{k} v_n^k x^k (1 - v_n x)^{n-k}, \qquad (12.24)$$

where $x \in [0, 1]$, $n \in \mathbb{N}$, $f \in AC([0, 1])$ with $f' \in L_\infty([0, 1])$. Since

$$L_n(e_0) = 1 + u_n,$$

we obtain that

$$st - \lim_n \|L_n(e_0) - e_0\| = st - \lim_n u_n = 0 = 0,$$

which implies (12.10). Also, by Hölder's inequality with $p = \dfrac{4}{3}$ and $q = 4$, since

$$L_n\left(|\Psi|^{\frac{3}{2}}; x\right) = (1 + u_n) \sum_{k=0}^{n} \left|x - \frac{k}{n}\right|^{3/2} \binom{n}{k} v_n^k x^k (1 - v_n x)^{n-k}$$

$$\leq (1 + u_n) \left(\sum_{k=0}^{n} \left(x - \frac{k}{n}\right)^2 \binom{n}{k} v_n^k x^k (1 - v_n x)^{n-k}\right)^{3/4}$$

$$= (1 + u_n) \left(\frac{v_n x - v_n^2 x^2}{n}\right)^{3/4},$$

we get

$$\delta_{n,\frac{1}{2}}^{3/2} = \left\| L_n\left(|\Psi|^{\frac{3}{2}}\right) \right\| \leq \frac{(1+u_n)}{(4n)^{3/4}}.$$

Using the fact that

$$st - \lim_n u_n = 0,$$

we obtain

$$st - \lim_n \frac{(1+u_n)}{(4n)^{3/4}} = 0.$$

Hence, we get

$$st - \lim_n \delta_{n,\frac{1}{2}} = 0,$$

which implies (12.11). Therefore, by Theorem 12.2, for every $f \in AC([0,1])$ with $f' \in L_\infty([0,1])$, we conclude that

$$st_A - \lim_n \|L_n(f) - f\| = 0.$$

However, since neither (u_n) nor (v_n) converges to zero (in the usual sense), it is impossible to approximate f by the sequence $(L_n(f))$ for every $f \in AC([0,1])$ with $f' \in L_\infty([0,1])$. Hence, this example demonstrates that the statistical approximation result in Theorem 12.2 is more applicable than Theorem A.

13
Fractional Trigonometric Korovkin Theory Based on Statistical Convergence

In this chapter, we develop the classical trigonometric Korovkin theory by using the concept of statistical convergence from the summability theory and also by considering the fractional derivatives of trigonometric functions. We also show that these results are more applicable than the classical ones. This chapter relies on [27].

13.1 Fractional Derivatives in Trigonometric Case

Throughout this chapter we focus on the closed interval $[-\pi, \pi]$. We now recall the Caputo fractional derivatives. Let r be a positive real number and $m = \lceil r \rceil$, where $\lceil \cdot \rceil$ is the ceiling of the number. Let $AC\left([-\pi, \pi]\right)$ denote the space of all real-valued absolutely continuous functions on $[-\pi, \pi]$. Consider the space

$$AC^m\left([-\pi, \pi]\right) := \left\{ f : [-\pi, \pi] \to \mathbb{R} : f^{(m-1)} \in AC\left([-\pi, \pi]\right) \right\}.$$

Then, the left Caputo fractional derivative of a function f belonging to $AC^m\left([-\pi, \pi]\right)$ is given by

$$D^r_{*(-\pi)} f(y) := \frac{1}{\Gamma(m-r)} \int\limits_{-\pi}^{y} (y-t)^{m-r-1} f^{(m)}(t) dt \quad \text{for} \quad y \in [-\pi, \pi], \quad (13.1)$$

G.A. Anastassiou and O. Duman: Towards Intelligent Modeling, ISRL 14, pp. 169–180.
springerlink.com   © Springer-Verlag Berlin Heidelberg 2011

where Γ is the usual Gamma function. Also, the right Caputo fractional derivative of a function f belonging to $AC^m\left([-\pi,\pi]\right)$ is defined by

$$D^r_{\pi-}f(y) := \frac{(-1)^m}{\Gamma(m-r)}\int_y^\pi (\zeta-y)^{m-r-1}f^{(m)}(\zeta)d\zeta \quad \text{for} \ \ y\in[-\pi,\pi]. \ (13.2)$$

In (13.1) and (13.2), we set $D^0_{*(-\pi)}f = f$ and $D^0_{\pi-}f = f$ on $[-\pi,\pi]$. We also consider the following assumptions:

$$D^r_{*(-\pi)}f(y) = 0 \ \text{for every} \ y < -\pi$$

and

$$D^r_{\pi-}f(y) = 0 \ \text{for every} \ y > \pi.$$

Then, the following facts are known (see, e.g., [13–15]):

(1°) If $r > 0$, $r \notin \mathbb{N}$, $m = \lceil r\rceil$, $f \in C^{m-1}([-\pi,\pi])$ and $f^{(m)} \in L_\infty\left([-\pi,\pi]\right)$, then we get $D^r_{*(-\pi)}f(-\pi) = 0$ and $D^r_{\pi-}f(\pi) = 0$.

(2°) Let $y \in [-\pi,\pi]$ be fixed. For $r > 0$, $m = \lceil r\rceil$, $f \in C^{m-1}([-\pi,\pi])$ with $f^{(m)} \in L_\infty\left([-\pi,\pi]\right)$, define the following Caputo fractional derivatives:

$$U_f(x,y) := D^r_{*x}f(y) = \frac{1}{\Gamma(m-r)}\int_x^y (y-t)^{m-r-1}f^{(m)}(t)dt \quad \text{for} \ \ y\in[x,\pi]$$

$$(13.3)$$

and

$$V_f(x,y) := D^r_{x-}f(y) = \frac{(-1)^m}{\Gamma(m-r)}\int_y^x (\zeta-y)^{m-r-1}f^{(m)}(\zeta)d\zeta \quad \text{for} \ \ y\in[-\pi,x].$$

$$(13.4)$$

Then, by [13, 15], for each fixed $x \in [-\pi,\pi]$, $U_f(x,.)$ is continuous on the interval $[x,\pi]$, and also $V_f(x,.)$ is continuous on $[-\pi,x]$. In addition, if $f \in C^m([-\pi,\pi])$, then, $U_f(\cdot,\cdot)$ and $V_f(\cdot,\cdot)$ are continuous on the set $[-\pi,\pi]\times[-\pi,\pi]$.

(3°) Let $\omega(f,\delta)$, $\delta > 0$, denote the usual modulus of continuity of a function f on $[-\pi,\pi]$. If $g \in C\left([-\pi,\pi]\times[-\pi,\pi]\right)$, then, for any $\delta > 0$, both the functions $s(x) := \omega\left(g\left(x,\cdot\right),\delta\right)_{[-\pi,x]}$ and $t(x) := \omega\left(g\left(x,\cdot\right),\delta\right)_{[x,\pi]}$ are continuous at the point $x \in [-\pi,\pi]]$.

($4°$) If $f \in C^{m-1}([-\pi, \pi])$ with $f^{(m)} \in L_\infty([-\pi, \pi])$, then we obtain from [15] that, for any $\delta > 0$,

$$\sup_{x \in [-\pi, \pi]} \omega\left(U_f(x, \cdot), \delta\right)_{[x, \pi]} < \infty \qquad (13.5)$$

and

$$\sup_{x \in [-\pi, \pi]} \omega\left(V_f(x, \cdot), \delta\right)_{[-\pi, x]} < \infty. \qquad (13.6)$$

($5°$) Now let $\Psi(y) := \Psi_x(y) = y - x$, $\Omega(y) := \Omega_x(y) = \sin\left(\frac{|y - x|}{4}\right)$ and $e_0(y) := 1$ on the interval $[-\pi, \pi]$. Following the paper by Anastassiou (see [15]) if $L_n : C([-\pi, \pi]) \to C([-\pi, \pi])$ is sequence of positive linear operators and if $r > 0$, $r \notin \mathbb{N}$, $m = \lceil r \rceil$, $f \in AC^m([-\pi, \pi])$ with $f^{(m)} \in L_\infty([-\pi, \pi])$, then we observe that ($\|.\|$ is the sup-norm)

$$\|L_n(f) - f\| \leq \|f\| \|L_n(e_0) - e_0\| + \sum_{k=1}^{m-1} \frac{\left\|f^{(k)}\right\|}{k!} \left\|L_n\left(|\Psi|^k\right)\right\|$$

$$+ \left(\frac{(2\pi)^r}{\Gamma(r+1)} \|L_n(e_0) - e_0\|^{\frac{1}{r+1}} + \frac{(2\pi)^r(r+1+2\pi)}{\Gamma(r+2)}\right)$$

$$\times \left\|L_n\left(\Omega^{r+1}\right)\right\|^{\frac{r}{r+1}} \left\{ \sup_{x \in [-\pi, \pi]} \omega\left(U_f(x, \cdot), \left\|L_n\left(\Omega^{r+1}\right)\right\|^{\frac{1}{r+1}}\right)_{[x, \pi]}\right.$$

$$\left. + \sup_{x \in [-\pi, \pi]} \omega\left(V_f(x, \cdot), \left\|L_n\left(\Omega^{r+1}\right)\right\|^{\frac{1}{r+1}}\right)_{[-\pi, x]}\right\}.$$

Then putting

$$\rho_{n,r} := \left\|L_n\left(\Omega^{r+1}\right)\right\|^{\frac{1}{r+1}}, \qquad (13.7)$$

and also considering (13.5), (13.6) we can write that

$$\|L_n(f) - f\| \leq K_{m,r} \left\{ \|L_n(e_0) - e_0\| + \sum_{k=1}^{m-1} \left\|L_n\left(|\Psi|^k\right)\right\|\right.$$

$$+ \rho_{n,r}^r \left(\sup_{x \in [-\pi, \pi]} \omega\left(U_f(x, \cdot), \rho_{n,r}\right)_{[x, \pi]}\right)$$

$$+ \rho_{n,r}^r \left(\sup_{x \in [-\pi, \pi]} \omega\left(V_f(x, \cdot), \rho_{n,r}\right)_{[-\pi, x]}\right)$$

$$+ \rho_{n,r}^r \|L_n(e_0) - e_0\|^{\frac{1}{r+1}} \left(\sup_{x \in [-\pi, \pi]} \omega\left(U_f(x, \cdot), \rho_{n,r}\right)_{[x, \pi]}\right)$$

$$\left. + \rho_{n,r}^r \|L_n(e_0) - e_0\|^{\frac{1}{r+1}} \left(\sup_{x \in [-\pi, \pi]} \omega\left(V_f(x, \cdot), \rho_{n,r}\right)_{[-\pi, x]}\right)\right\},$$

$$(13.8)$$

where

$$K_{r,m} := \max \left\{ \frac{(2\pi)^r}{\Gamma(r+1)}, \frac{(2\pi)^r (r+1+2\pi)}{\Gamma(r+2)}, \|f\|, \|f'\|, \frac{\|f''\|}{2!}, \frac{\|f'''\|}{3!}, ..., \frac{\left\|f^{(m-1)}\right\|}{(m-1)!} \right\}.$$

(13.9)

We should note that the sum in the right hand-side of (13.8) collapses when $r \in (0, 1)$.

Hence, the following theorem is a fractional Korovkin-type approximation result for a sequence of positive linear operators.

Theorem A (see [15]). *Let $L_n : C([-\pi, \pi]) \to C([-\pi, \pi])$ be a sequence of positive linear operators, and let $r > 0$, $r \notin \mathbb{N}$, $m = \lceil r \rceil$. If the sequence $(\rho_{n,r})$ given by (13.7) is convergent to zero as n tends to infinity and $(L_n(e_0))$ converges uniformly to e_0 on $[-\pi, \pi]$, then, for every $f \in AC^m([-\pi, \pi])$ with $f^{(m)} \in L_\infty([-\pi, \pi])$, the sequence $(L_n(f))$ converges uniformly to f on the interval $[-\pi, \pi]$. Furthermore, this uniform convergence is still valid on $[-\pi, \pi]$ when $f \in C^m([-\pi, \pi])$.*

13.2 Fractional Trigonometric Korovkin Results in Statistical Sense

In this section, we get the statistical version of Theorem A. We first need the next lemma.

Lemma 13.1. *Let $A = [a_{jn}]$ be a non-negative regular summability matrix, and let $r > 0$, $r \notin \mathbb{N}$, $m = \lceil r \rceil$. Assume that $L_n : C([-\pi, \pi]) \to C([-\pi, \pi])$ is a sequence of positive linear operators. If*

$$st_A - \lim_n \|L_n(e_0) - e_0\| = 0 \tag{13.10}$$

and

$$st_A - \lim_n \rho_{n,r} = 0, \tag{13.11}$$

where $\rho_{n,r}$ is the same as in (13.7), then we get, for every $k = 1, 2, ..., m-1$,

$$st_A - \lim_n \left\| L_n\left(|\Psi|^k\right) \right\| = 0.$$

Proof. Let $k \in \{1, 2, ..., m-1\}$ be fixed. Then, applying Hölder's inequality for positive linear operators with $p = \dfrac{r+1}{k}$, $q = \dfrac{r+1}{r+1-k}$ $\left(\dfrac{1}{p} + \dfrac{1}{q} = 1\right)$, we see that

$$\left\| L_n \left(|\Psi|^k \right) \right\| = 2^k \left\| L_n \left(\left(\frac{|\Psi|}{2} \right)^k \right) \right\|$$

$$\leq 2^k \left\| L_n \left(\left(\frac{|\Psi|}{2} \right)^{r+1} \right) \right\|^{\frac{k}{r+1}} \left\| L_n (e_0) \right\|^{\frac{r+1-k}{r+1}},$$

which implies

$$\left\| L_n \left(|\Psi|^k \right) \right\| \leq 2^k \left\| L_n \left(\left(\frac{|\Psi|}{2} \right)^{r+1} \right) \right\|^{\frac{k}{r+1}} \left\{ \left\| L_n (e_0) - e_0 \right\|^{\frac{r+1-k}{r+1}} + 1 \right\}.$$

Now considering the fact that $|u| \leq \pi \sin (|u|/2)$ for $u \in [-\pi, \pi]$, we get

$$\left\| L_n \left(|\Psi|^k \right) \right\| \leq (2\pi)^k \left\| L_n \left(\sin^{r+1} \left(\frac{|\Psi|}{4} \right) \right) \right\|^{\frac{k}{r+1}} \left\{ \left\| L_n (e_0) - e_0 \right\|^{\frac{r+1-k}{r+1}} + 1 \right\}.$$

Hence, for each $k = 1, 2, ..., m - 1$, we obtain the following inequality:

$$\left\| L_n \left(|\Psi|^k \right) \right\| \leq (2\pi)^k \left(\rho_{n,r}^k \left\| L_n (e_0) - e_0 \right\|^{\frac{r+1-k}{r+1}} + \rho_{n,r}^k \right). \qquad (13.12)$$

Then, for a given $\varepsilon > 0$, consider the following sets:

$$A := \left\{ n \in \mathbb{N} : \left\| L_n \left(|\Psi|^k \right) \right\| \geq \varepsilon \right\},$$

$$A_1 := \left\{ n \in \mathbb{N} : \rho_{n,r}^k \left\| L_n (e_0) - e_0 \right\|^{\frac{r+1-k}{r+1}} \geq \frac{\varepsilon}{2 (2\pi)^k} \right\}$$

$$A_2 := \left\{ n \in \mathbb{N} : \rho_{n,r} \geq \frac{1}{2\pi} \left(\frac{\varepsilon}{2} \right)^{\frac{1}{k}} \right\}.$$

Then, it follows from (13.12) that $A \subseteq A_1 \cup A_2$. Also, considering

$$A_1' := \left\{ n \in \mathbb{N} : \rho_{n,r} \geq \frac{1}{\sqrt{2\pi}} \left(\frac{\varepsilon}{2} \right)^{\frac{1}{2k}} \right\},$$

$$A_1'' := \left\{ n \in \mathbb{N} : \left\| L_n (e_0) - e_0 \right\| \geq \left(\frac{\varepsilon}{2 (2\pi)^k} \right)^{\frac{r+1}{2(r+1-k)}} \right\},$$

we see that $A_1 \subseteq A_1' \cup A_1''$, which gives

$$A \subseteq A_1' \cup A_1'' \cup A_2.$$

Hence, for every $j \in \mathbb{N}$, we obtain

$$\sum_{n \in A} a_{jn} \leq \sum_{n \in A_1'} a_{jn} + \sum_{n \in A_1''} a_{jn} + \sum_{n \in A_2} a_{jn}.$$

Taking $j \to \infty$ in the last inequality and also using the hypotheses (13.10) and (13.11) we immediately observe that

$$\lim_{j} \sum_{n \in A} a_{jn} = 0.$$

Hence, we deduce that, for each $k = 1, 2, ..., m - 1$,

$$st_A - \lim_{n} \left\| L_n \left(|\Psi|^k \right) \right\| = 0,$$

whence the result.   ∎

Then we obtain the following fractional approximation result based on statistical convergence.

Theorem 13.2. *Let* $A = [a_{jn}]$ *be a non-negative regular summability matrix, and let* $r > 0$, $r \notin \mathbb{N}$, $m = \lceil r \rceil$. *Assume that* $L_n : C([-\pi, \pi]) \to C([-\pi, \pi])$ *is a sequence of positive linear operators. If* (13.10) *and* (13.11) *hold, then, for every* $f \in AC^m([-\pi, \pi])$ *with* $f^{(m)} \in L_\infty([-\pi, \pi])$, *we get*

$$st_A - \lim_{n} \|L_n(f) - f\| = 0. \tag{13.13}$$

Proof. Let $f \in AC^m([-\pi, \pi])$ with $f^{(m)} \in L_\infty([-\pi, \pi])$. Then, using (13.5), (13.6) and (13.8), we have

$$\|L_n(f) - f\| \leq M_{m,r} \left\{ \|L_n(e_0) - e_0\| + 2\rho_{n,r}^r \right.$$
$$\left. + 2\rho_{n,r}^r \|L_n(e_0) - e_0\|^{\frac{1}{r+1}} + \sum_{k=1}^{m-1} \left\| L_n \left(|\Psi|^k \right) \right\| \right\}, \tag{13.14}$$

where

$$M_{m,r} := \max \left\{ K_{m,r}, \sup_{x \in [-\pi, \pi]} \omega \left(U_f(x, \cdot), \rho_{n,r} \right)_{[x, \pi]}, \sup_{x \in [-\pi, \pi]} \omega \left(V_f(x, \cdot), \rho_{n,r} \right)_{[-\pi, x]} \right\}$$

and $K_{m,r}$ is given by (13.9). Now, for a given $\varepsilon > 0$, consider the following sets:

$$B := \left\{ n \in \mathbb{N} : \|L_n(f) - f\| \geq \varepsilon \right\},$$

$$B_k := \left\{ n \in \mathbb{N} : \left\| L_n \left(|\Psi|^k \right) \right\| \geq \frac{\varepsilon}{(m+2)M_{m,r}} \right\}, \quad k = 1, 2, ..., m - 1,$$

$$B_m := \left\{ n \in \mathbb{N} : \|L_n(e_0) - e_0\| \geq \frac{\varepsilon}{(m+2)M_{m,r}} \right\},$$

$$B_{m+1} := \left\{ n \in \mathbb{N} : \delta_{n,r} \geq \left(\frac{\varepsilon}{2(m+2)M_{m,r}} \right)^{\frac{1}{r}} \right\},$$

$$B_{m+2} := \left\{ n \in \mathbb{N} : \delta_{n,r}^r \|L_n(e_0) - e_0\|^{\frac{1}{r+1}} \geq \frac{\varepsilon}{2(m+2)M_{m,r}} \right\}.$$

Then, it follows from (13.14) that $B \subseteq \overset{m+2}{\underset{i=1}{\bigcup}} B_i$. Also considering

$$B_{m+3} := \left\{ n \in \mathbb{N} : \|L_n(e_0) - e_0\| \geq \left(\frac{\varepsilon}{2(m+2)M_{m,r}} \right)^{\frac{r+1}{2}} \right\}$$

and

$$B_{m+4} := \left\{ n \in \mathbb{N} : \delta_{n,r} \geq \left(\frac{\varepsilon}{2(m+2)M_{m,r}} \right)^{\frac{1}{2r}} \right\},$$

we observe that

$$B_{m+2} \subseteq B_{m+3} \cup B_{m+4},$$

which gives

$$B \subseteq \overset{m+4}{\underset{\substack{i=1 \ (i \neq m+2)}}{\bigcup}} B_i.$$

Hence, for every $j \in \mathbb{N}$, we have

$$\sum_{n \in B} a_{jn} \leq \sum_{\substack{i=1 \ (i \neq m+2)}}^{m+4} \sum_{n \in B_i} a_{jn}. \tag{13.15}$$

Letting $j \to \infty$ in the both sides of (13.15) and also using (13.10), (13.11), and also considering Lemma 13.1 we deduce that

$$\lim_j \sum_{n \in B} a_{jn} = 0,$$

which implies (13.13). ∎

If we use the space $C^m([-\pi, \pi])$ instead of $AC^m([-\pi, \pi])$, then we can obtain a slight modification of Theorem 13.2. To see this we need the following lemma.

Lemma 13.3. *Let $A = [a_{jn}]$ be a non-negative regular summability matrix, and let $r > 0$, $r \notin \mathbb{N}$, $m = \lceil r \rceil$. Assume that $L_n : C([-\pi, \pi]) \to C([-\pi, \pi])$ is a sequence of positive linear operators. If (13.11) holds, then, for every $f \in C^m([-\pi, \pi])$, we get:*

(i) $st_A - \underset{n}{\lim} \left(\underset{x \in [-\pi, \pi]}{\sup} \omega \left(U_f(x, \cdot), \rho_{n,r} \right)_{[x, \pi]} \right) = 0,$

(ii) $st_A - \underset{n}{\lim} \left(\underset{x \in [-\pi, \pi]}{\sup} \omega \left(V_f(x, \cdot), \rho_{n,r} \right)_{[-\pi, x]} \right) = 0,$

where $\rho_{n,r}$ is the same as in (13.7); $U_f(\cdot, \cdot)$ and $V_f(\cdot, \cdot)$ are given respectively by (13.3) and (13.4).

Proof. From (**2°**), if $f \in C^m([-\pi, \pi])$, then both $U_f(\cdot, \cdot)$ and $V_f(\cdot, \cdot)$ belong to $C([-\pi, \pi] \times [-\pi, \pi])$. Then, by (**3°**), the functions $\omega(U_f(x, \cdot), \delta_{n,r})_{[x,\pi]}$ and $\omega(V_f(x, \cdot), \delta_{n,r})_{[-\pi,x]}$ are continuous at the point $x \in [-\pi, \pi]$. Hence, there exist the points $x_0, x_1 \in [-\pi, \pi]$ such that

$$\sup_{x \in [-\pi, \pi]} \omega(U_f(x, \cdot), \rho_{n,r})_{[x,\pi]} = \omega(U_f(x_0, \cdot), \rho_{n,r})_{[x_0,\pi]} =: g(\rho_{n,r})$$

and

$$\sup_{x \in [-\pi, \pi]} \omega(V_f(x, \cdot), \rho_{n,r})_{[-\pi,x]} = \omega(V_f(x_1, \cdot), \rho_{n,r})_{[-\pi,x_1]} =: h(\rho_{n,r}).$$

Since $U_f(x_0, \cdot)$ and $V_f(x_1, \cdot)$ are continuous on $[-\pi, \pi]$, the functions g and h are right continuous at the origin. By (13.11), we obtain, for any $\delta > 0$, that

$$\lim_j \sum_{n:\rho_{n,r} \geq \delta} a_{jn} = 0. \tag{13.16}$$

Now, by the right continuity of g and h at zero, for a given $\varepsilon > 0$, there exist $\delta_1, \delta_2 > 0$ such that $g(\delta_{n,r}) < \varepsilon$ whenever $\delta_{n,r} < \delta_1$ and that $h(\delta_{n,r}) < \varepsilon$ whenever $\delta_{n,r} < \delta_2$. Then, we can write that $g(\delta_{n,r}) \geq \varepsilon$ gives $\delta_{n,r} \geq \delta_1$, and also that $h(\delta_{n,r}) \geq \varepsilon$ gives $\delta_{n,r} \geq \delta_2$. Hence, we observe that

$$\{n \in \mathbb{N} : g(\rho_{n,r}) \geq \varepsilon\} \subseteq \{n \in \mathbb{N} : \rho_{n,r} \geq \delta_1\} \tag{13.17}$$

and

$$\{n \in \mathbb{N} : h(\rho_{n,r}) \geq \varepsilon\} \subseteq \{n \in \mathbb{N} : \rho_{n,r} \geq \delta_2\} \tag{13.18}$$

So, it follows from (13.17) and (13.18) that, for each $j \in \mathbb{N}$,

$$\sum_{n:g(\rho_{n,r}) \geq \varepsilon} a_{jn} \leq \sum_{n:\rho_{n,r} \geq \delta_1} a_{jn} \tag{13.19}$$

and

$$\sum_{n:h(\rho_{n,r}) \geq \varepsilon} a_{jn} \leq \sum_{n:\rho_{n,r} \geq \delta_2} a_{jn} \tag{13.20}$$

Then, letting $j \to \infty$ on the both sides of the inequalities (13.19), (13.20); and also using (13.16) we immediately see, for every $\varepsilon > 0$, that

$$\lim_j \sum_{n:g(\rho_{n,r}) \geq \varepsilon} a_{jn} = \lim_j \sum_{n:h(\rho_{n,r}) \geq \varepsilon} a_{jn} = 0,$$

which implies that

$$st_A - \lim_n \left(\sup_{x \in [-\pi, \pi]} \omega(U_f(x, \cdot), \rho_{n,r})_{[x,\pi]} \right) = 0$$

and

$$st_A - \lim_n \left(\sup_{x \in [-\pi, \pi]} \omega \left(V_f(x, \cdot), \rho_{n,r} \right)_{[-\pi, x]} \right) = 0.$$

Therefore, the proof of Lemma is done. ∎

Then, we obtain the next result.

Theorem 13.4. *Let $A = [a_{jn}]$ be a non-negative regular summability matrix, and let $r > 0$, $r \notin \mathbb{N}$, $m = \lceil r \rceil$. Assume that $L_n : C([-\pi, \pi]) \to C([-\pi, \pi])$ is a sequence of positive linear operators. If (13.10) and (13.11) hold, then, for every $f \in C^m([-\pi, \pi])$, we get (13.13).*

Proof. By (13.8), we have

$$
\begin{aligned}
\|L_n(f) - f\| \leq K_{m,r} \Big\{ & \|L_n(e_0) - e_0\| + \sum_{k=1}^{m-1} \left\| L_n\left(|\Psi|^k\right) \right\| \\
& + \rho_{n,r}^r g\left(\rho_{n,r}\right) + \rho_{n,r}^r h\left(\rho_{n,r}\right) \\
& + \rho_{n,r}^r g\left(\rho_{n,r}\right) \|L_n(e_0) - e_0\|^{\frac{1}{r+1}} \\
& + \rho_{n,r}^r h\left(\rho_{n,r}\right) \|L_n(e_0) - e_0\|^{\frac{1}{r+1}} \Big\},
\end{aligned}
\tag{13.21}
$$

where $g(\rho_{n,r})$ and $h(\rho_{n,r})$ are the same as in the proof of Lemma 13.3. Now, for a given $\varepsilon > 0$, define the following sets

$$C := \{n \in \mathbb{N} : \|L_n(f) - f\| \geq \varepsilon\},$$

$$C_k := \left\{ n \in \mathbb{N} : \left\| L_n\left(|\Psi|^k\right) \right\| \geq \frac{\varepsilon}{(m+4)K_{m,r}} \right\}, \quad k = 1, 2, ..., m-1.$$

$$C_m := \left\{ n \in \mathbb{N} : \|L_n(e_0) - e_0\| \geq \frac{\varepsilon}{(m+4)K_{m,r}} \right\}$$

$$C_{m+1} := \left\{ n \in \mathbb{N} : \rho_{n,r}^r g\left(\rho_{n,r}\right) \geq \frac{\varepsilon}{(m+4)K_{m,r}} \right\},$$

$$C_{m+2} := \left\{ n \in \mathbb{N} : \rho_{n,r}^r h\left(\rho_{n,r}\right) \geq \frac{\varepsilon}{(m+4)K_{m,r}} \right\},$$

$$C_{m+3} := \left\{ n \in \mathbb{N} : \rho_{n,r}^r g\left(\rho_{n,r}\right) \|L_n(e_0) - e_0\|^{\frac{1}{r+1}} \geq \frac{\varepsilon}{(m+4)K_{m,r}} \right\}$$

$$C_{m+4} := \left\{ n \in \mathbb{N} : \rho_{n,r}^r h\left(\rho_{n,r}\right) \|L_n(e_0) - e_0\|^{\frac{1}{r+1}} \geq \frac{\varepsilon}{(m+4)K_{m,r}} \right\}.$$

Then, by (13.21), we get

$$C \subseteq \bigcup_{i=1}^{m+4} C_i.$$

So, for every $j \in \mathbb{N}$, we observe that

$$\sum_{n \in C} a_{jn} \leq \sum_{i=1}^{m+4} \left(\sum_{n \in C_i} a_{jn} \right). \tag{13.22}$$

On the other hand, by (13.10), (13.11) and Lemmas 13.1, 13.3, we derive that

$$st_A - \lim_n \left\| L_n \left(|\Psi|^k \right) \right\| = 0, \ (k = 1, .., m - 1),$$

$$st_A - \lim_n \rho_{n,r}^r g \left(\rho_{n,r} \right) = 0,$$

$$st_A - \lim_n \rho_{n,r}^r h \left(\rho_{n,r} \right) = 0,$$

$$st_A - \lim_n \rho_{n,r}^r g \left(\rho_{n,r} \right) \| L_n \left(e_0 \right) - e_0 \|^{\frac{1}{r+1}} = 0,$$

$$st_A - \lim_n \rho_{n,r}^r h \left(\rho_{n,r} \right) \| L_n \left(e_0 \right) - e_0 \|^{\frac{1}{r+1}} = 0.$$

Hence, we see that, for every $i = 1, 2, ..., m + 4$,

$$\lim_j \sum_{n \in C_i} a_{jn} = 0. \tag{13.23}$$

Now, taking limit as $j \to \infty$ in the both sides of (13.22) and using (13.23) we conclude that

$$\lim_j \sum_{n \in C} a_{jn} = 0.$$

The last equality gives that

$$st_A - \lim_n \| L_n(f) - f \| = 0,$$

which finishes the proof.   ∎

13.3 Conclusions

In this section we give a sequence of positive linear operators which satisfies all conditions of Theorem 13.2 but not Theorem A.

Now take $A = C_1 = [c_{jn}]$, the Cesáro matrix, and define the sequences (u_n) and (v_n) by

$$u_n := \begin{cases} \sqrt{n}, & \text{if } n = m^2 \ (m \in \mathbb{N}), \\ \frac{1}{n}, & \text{otherwise.} \end{cases}$$

and

$$v_n := \begin{cases} 1/2, & \text{if } n = m^2 \ (m \in \mathbb{N}), \\ \frac{n}{n+1}, & \text{otherwise.} \end{cases}$$

Then observe that

$$st - \lim_n u_n = 0 \quad \text{and} \quad st - \lim_n v_n = 1. \tag{13.24}$$

Let $r = \dfrac{1}{2}$. Then we get $m = \left[\dfrac{1}{2}\right] = 1$. Now define the following Bernstein-like positive linear operators:

$$L_n(f;x) := (1+u_n) \sum_{k=0}^{n} f\left(\frac{2\pi k}{n} - \pi\right) \binom{n}{k} \left(\frac{\pi + v_n x}{2\pi}\right)^k \left(\frac{\pi - v_n x}{2\pi}\right)^{n-k},$$

$$(13.25)$$

where $x \in [-\pi, \pi]$, $n \in \mathbb{N}$, $f \in AC([-\pi,\pi])$ with $f' \in L_\infty([-\pi,\pi])$. Since

$$L_n(e_0) = 1 + u_n,$$

we easily obtain that

$$st - \lim_n \|L_n(e_0) - e_0\| = st - \lim_n u_n = 0,$$

which implies (13.10). Also, by Hölder's inequality with $p = \dfrac{4}{3}$ and $q = 4$, since, for every $x \in [-\pi, \pi]$,

$$L_n\left(|\Psi|^{\frac{3}{2}};x\right) = (1+u_n) \sum_{k=0}^{n} \left|x + \pi - \frac{2\pi k}{n}\right|^{3/2} \binom{n}{k} \left(\frac{\pi + v_n x}{2\pi}\right)^k \left(\frac{\pi - v_n x}{2\pi}\right)^{n-k}$$

$$\leq (1+u_n)\left(\sum_{k=0}^{n}\left(x + \pi - \frac{2\pi k}{n}\right)^2 \binom{n}{k} \left(\frac{\pi + v_n x}{2\pi}\right)^k \left(\frac{\pi - v_n x}{2\pi}\right)^{n-k}\right)^{3/4}$$

$$= (1+u_n)\left(x^2(1-v_n)^2 + \frac{\pi^2 - v_n^2 x^2}{n}\right)^{3/4},$$

we get

$$\left\|L_n\left(|\Psi|^{\frac{3}{2}}\right)\right\| \leq \pi^{3/2}(1+u_n)\left((1-v_n)^2 + \frac{1}{n}\right)^{3/4}.$$

$$(13.26)$$

Since $|\sin u| \leq |u|$, it follows from (13.26) that

$$\rho_{n,\frac{1}{2}}^{3/2} = \left\|L_n\left(\Omega^{\frac{3}{2}}\right)\right\| \leq \frac{1}{8}\left\|L_n\left(|\Psi|^{\frac{3}{2}}\right)\right\| \leq \frac{\pi^{3/2}(1+u_n)}{8}\left((1-v_n)^2 + \frac{1}{n}\right)^{3/4}.$$

$$(13.27)$$

Now using (13.24), we see that

$$st - \lim_n \frac{\pi^{3/2}(1+u_n)}{8}\left((1-v_n)^2 + \frac{1}{n}\right)^{3/4} = 0.$$

Hence, we get from (13.27) that

$$st - \lim_n \rho_{n,\frac{1}{2}} = 0,$$

which implies (13.11). Thus, by Theorem 13.2, for every $f \in AC([-\pi, \pi])$ with $f' \in L_\infty([-\pi, \pi])$, we get

$$st_A - \lim_n \|L_n(f) - f\| = 0.$$

However, since neither (u_n) nor (v_n) converges to zero (in the usual sense), it is impossible to approximate f by the sequence $(L_n(f))$ for every $f \in AC([-\pi, \pi])$ with $f' \in L_\infty([-\pi, \pi])$. This example clearly gives us that the statistical result in Theorem 13.2 is more applicable than Theorem A.

14

Statistical Fuzzy Approximation Theory by Fuzzy Positive Linear Operators

In this chapter, we give a Korovkin-type approximation theorem for fuzzy positive linear operators by using the notion of A-statistical convergence, where A is a non-negative regular summability matrix. This type of approximation enables us to obtain more powerful results than in the classical aspects of approximation theory settings. An application of this result is also presented. Furthermore, we study the rates of this statistical fuzzy convergence of the operators via the fuzzy modulus of continuity. This chapter relies on [17].

14.1 Statistical Fuzzy Korovkin Theory

In this section we get a fuzzy Korovkin-type theorem via the concept of A-statistical convergence. Also, we present an example of fuzzy positive linear operators by using fuzzy Bernstein polynomials, which indicates that the results obtained in this section are stronger than the classical case.

Let $f : [a, b] \to \mathbb{R}_{\mathcal{F}}$ be fuzzy number valued functions. Then f is said to be fuzzy continuous at $x_0 \in [a, b]$ provided that whenever $x_n \to x_0$, then $D\left(f(x_n), f(x_0)\right) \to \infty$ as $n \to \infty$. Also, we say that f is fuzzy continuous on $[a, b]$ if it is fuzzy continuous at every point $x \in [a, b]$. The set of all fuzzy continuous functions on the interval $[a, b]$ is denoted by $C_{\mathcal{F}}[a, b]$ (see, for instance, [6]). Notice that $C_{\mathcal{F}}[a, b]$ is only a cone not a vector space. Now let $L : C_{\mathcal{F}}[a, b] \to C_{\mathcal{F}}[a, b]$ be an operator. Then L is said to be fuzzy linear if, for every $\lambda_1, \lambda_2 \in \mathbb{R}$, $f_1, f_2 \in C_{\mathcal{F}}[a, b]$, and $x \in [a, b]$,

G.A. Anastassiou and O. Duman: Towards Intelligent Modeling, ISRL 14, pp. 181–188.
springerlink.com © Springer-Verlag Berlin Heidelberg 2011

$$L\left(\lambda_1 \odot f_1 \oplus \lambda_2 \odot f_2; x\right) = \lambda_1 \odot L(f_1; x) \oplus \lambda_2 \odot L(f_2; x)$$

holds. Also L is called fuzzy positive linear operator if it is fuzzy linear and, the condition $L(f; x) \preceq L(g; x)$ is satisfied for any $f, g \in C_{\mathcal{F}}[a, b]$ and all $x \in [a, b]$ with $f(x) \preceq g(x)$.

Here we consider the test functions e_i given by $e_i(x) = x^i$, $i = 0, 1, 2$. Then, in [6], the next Korovkin theorem was proved.

Theorem A (see [6]). *Let (L_n) be a sequence of fuzzy positive linear operators from $C_{\mathcal{F}}[a, b]$ into itself. Assume that there exists a corresponding sequence $(\tilde{L}_n)$ of positive linear operators from $C[a, b]$ into itself with the property*

$$\{L_n(f; x)\}_{\pm}^{(r)} = \tilde{L}_n\left(f_{\pm}^{(r)}; x\right) \tag{14.1}$$

for all $x \in [a, b]$, $r \in [0, 1]$, $n \in \mathbb{N}$ and $f \in C_{\mathcal{F}}[a, b]$. Assume further that

$$\lim_n \left\|\tilde{L}_n(e_i) - e_i\right\| = 0 \text{ for each } i = 0, 1, 2.$$

Then, for all $f \in C_{\mathcal{F}}[a, b]$, we get

$$\lim_n D^*\left(L_n(f), f\right) = 0.$$

Then, we obtain the next result.

Theorem 14.1. *Let $A = [a_{jn}]$ be a non-negative regular summability matrix and let (L_n) be a sequence of fuzzy positive linear operators from $C_{\mathcal{F}}[a, b]$ into itself. Assume that there exists a corresponding sequence $(\tilde{L}_n)$ of positive linear operators from $C[a, b]$ into itself with the property (14.1). Assume further that*

$$st_A - \lim_n \left\|\tilde{L}_n(e_i) - e_i\right\| = 0 \text{ for each } i = 0, 1, 2. \tag{14.2}$$

Then, for all $f \in C_{\mathcal{F}}[a, b]$, we get

$$st_A - \lim_n D^*\left(L_n(f), f\right) = 0.$$

Proof. Let $f \in C_{\mathcal{F}}[a, b]$, $x \in [a, b]$ and $r \in [0, 1]$. By the hypothesis, since $f_{\pm}^{(r)} \in C[a, b]$, we can write, for every $\varepsilon > 0$, that there exists a number $\delta > 0$ such that $\left|f_{\pm}^{(r)}(y) - f_{\pm}^{(r)}(x)\right| < \varepsilon$ holds for every $y \in [a, b]$ satisfying $|y - x| < \delta$. Then we immediately see, for all $y \in [a, b]$, that

$$\left|f_{\pm}^{(r)}(y) - f_{\pm}^{(r)}(x)\right| \le \varepsilon + 2M_{\pm}^{(r)} \frac{(y - x)^2}{\delta^2},$$

where $M_{\pm}^{(r)} := \left\|f_{\pm}^{(r)}\right\|$. Now using the linearity and the positivity of the operators $\tilde{L}_n$, we get, for each $n \in \mathbb{N}$, that

$$\left|\tilde{L}_n\left(f_{\pm}^{(r)};x\right) - f_{\pm}^{(r)}(x)\right| \leq \tilde{L}_n\left(\left|f_{\pm}^{(r)}(y) - f_{\pm}^{(r)}(x)\right|;x\right)$$

$$+M_{\pm}^{(r)}\left|\tilde{L}_n\left(e_0;x\right) - e_0(x)\right|$$

$$\leq \varepsilon + \left(\varepsilon + M_{\pm}^{(r)}\right)\left|\tilde{L}_n\left(e_0;x\right) - e_0(x)\right|$$

$$+\frac{2M_{\pm}^{(r)}}{\delta^2}\left|\tilde{L}_n\left((y-x)^2;x\right)\right|$$

which implies

$$\left|\tilde{L}_n\left(f_{\pm}^{(r)};x\right) - f_{\pm}^{(r)}(x)\right| \leq \varepsilon + \left(\varepsilon + M_{\pm}^{(r)} + \frac{2c^2M_{\pm}^{(r)}}{\delta^2}\right)\left|\tilde{L}_n\left(e_0;x\right) - e_0(x)\right|$$

$$+\frac{4cM_{\pm}^{(r)}}{\delta^2}\left|\tilde{L}_n\left(e_1;x\right) - e_1(x)\right|$$

$$+\frac{2M_{\pm}^{(r)}}{\delta^2}\left|\tilde{L}_n\left(e_2;x\right) - e_2(x)\right|,$$

where $c := \max\{|a|, |b|\}$. Also defining

$$K_{\pm}^{(r)}(\varepsilon) := \max\left\{\varepsilon + M_{\pm}^{(r)} + \frac{2c^2M_{\pm}^{(r)}}{\delta^2}, \frac{4cM_{\pm}^{(r)}}{\delta^2}, \frac{2M_{\pm}^{(r)}}{\delta^2}\right\}$$

and taking supremum over $x \in [a,b]$, the above inequality gives that

$$\left\|\tilde{L}_n\left(f_{\pm}^{(r)}\right) - f_{\pm}^{(r)}\right\| \leq \varepsilon + K_{\pm}^{(r)}(\varepsilon)\{\left\|\tilde{L}_n\left(e_0\right) - e_0\right\|$$
$$+ \left\|\tilde{L}_n\left(e_1\right) - e_1\right\| + \left\|\tilde{L}_n\left(e_2\right) - e_2\right\|\}. \tag{14.3}$$

Now it follows from (14.1) that

$$D^*\left(L_n(f), f\right) = \sup_{x \in [a,b]} D\left(L_n(f;x), f(x)\right)$$

$$= \sup_{x \in [a,b]} \sup_{r \in [0,1]} \max\left\{\left|\tilde{L}_n\left(f_-^{(r)};x\right) - f_-^{(r)}(x)\right|, \left|\tilde{L}_n\left(f_+^{(r)};x\right) - f_+^{(r)}(x)\right|\right\}$$

$$= \sup_{r \in [0,1]} \max\left\{\left\|\tilde{L}_n\left(f_-^{(r)}\right) - f_-^{(r)}\right\|, \left\|\tilde{L}_n\left(f_+^{(r)}\right) - f_+^{(r)}\right\|\right\}.$$

Combining the above equality with (14.3), we get

$$D^*\left(L_n(f), f\right) \leq \varepsilon + K(\varepsilon)\{\left\|\tilde{L}_n\left(e_0\right) - e_0\right\| + \left\|\tilde{L}_n\left(e_1\right) - e_1\right\|$$
$$+ \left\|\tilde{L}_n\left(e_2\right) - e_2\right\|\}, \tag{14.4}$$

where $K(\varepsilon) := \sup_{r \in [0,1]} \max\left\{K_-^{(r)}(\varepsilon), K_+^{(r)}(\varepsilon)\right\}$. Now, for a given $\varepsilon' > 0$, chose $\varepsilon > 0$ such that $0 < \varepsilon < \varepsilon'$, and also consider the following sets:

$$U := \left\{n \in \mathbb{N} : D^*\left(L_n(f), f\right) \geq \varepsilon'\right\},$$

$$U_0 := \left\{n \in \mathbb{N} : \left\|\tilde{L}_n\left(e_0\right) - e_0\right\| \geq \frac{\varepsilon' - \varepsilon}{3K(\varepsilon)}\right\},$$

$$U_1 := \left\{n \in \mathbb{N} : \left\|\tilde{L}_n\left(e_1\right) - e_1\right\| \geq \frac{\varepsilon' - \varepsilon}{3K(\varepsilon)}\right\},$$

$$U_2 := \left\{n \in \mathbb{N} : \left\|\tilde{L}_n\left(e_2\right) - e_2\right\| \geq \frac{\varepsilon' - \varepsilon}{3K(\varepsilon)}\right\}.$$

Then inequality (14.4) yields

$$U \subseteq U_0 \cup U_1 \cup U_2,$$

which implies that, for each $j \in \mathbb{N}$,

$$\sum_{n \in U} a_{jn} \leq \sum_{n \in U_0} a_{jn} + \sum_{n \in U_1} a_{jn} + \sum_{n \in U_2} a_{jn}. \tag{14.5}$$

If we take limit as $j \to \infty$ on the both sides of inequality (14.5) and use the hypothesis (14.2), we immediately obtain that

$$\lim_j \sum_{n \in U} a_{jn} = 0,$$

which completes the proof. ∎

Remark 14.2. *If we replace the matrix A in Theorem 14.1 by the identity matrix, then we obtain Theorem A given by Anastassiou in [6] at once. However, we can construct a sequence of fuzzy positive linear operators, which satisfies the statistical fuzzy approximation result (Theorem 14.1), but not Theorem A.*

Take $A = C_1 = [c_{jn}]$, the Cesáro matrix of order one and define the sequence (u_n) by:

$$u_n = \begin{cases} 1, & \text{if } n \neq m^2, \ (m = 1, 2, \ldots), \\ \sqrt{n}, & \text{otherwise}. \end{cases} \tag{14.6}$$

Then define the fuzzy Bernstein-type polynomials by

$$B_n^{\mathcal{F}}(f; x) = u_n \odot \bigoplus_{k=0}^{n} \binom{n}{k} x^k (1 - x)^{n-k} \odot f\left(\frac{k}{n}\right),$$

where $f \in C_{\mathcal{F}}[0, 1]$, $x \in [0, 1]$ and $n \in \mathbb{N}$. In this case, we see that

$$\left\{B_n^{\mathcal{F}}(f; x)\right\}_{\pm}^{(r)} = \tilde{B}_n\left(f_{\pm}^{(r)}; x\right) = u_n \sum_{k=0}^{n} \binom{n}{k} x^k (1 - x)^{n-k} f_{\pm}^{(r)}\left(\frac{k}{n}\right),$$

where $f_{\pm}^{(r)} \in C[0, 1]$. It is easy to check that

$$\tilde{B}_n \left(e_0; x \right) = u_n,$$
$$\tilde{B}_n \left(e_1; x \right) = x u_n,$$
$$\tilde{B}_n \left(e_2; x \right) = \left(x^2 + \frac{x(1-x)}{n} \right) u_n.$$

Since

$$\sum_{n:|u_n-1|\geq\varepsilon} c_{jn} = \sum_{n:|u_n-1|\geq\varepsilon} \frac{1}{j} \leq \frac{\sqrt{j}}{j} = \frac{1}{\sqrt{j}} \to 0 \ \ (\text{as } j \to \infty),$$

we have

$$st_{C_1} - \lim_n u_n = 1.$$

The above gives that

$$st_{C_1} - \lim_n \left\| \tilde{B}_n(e_i) - e_i \right\| = 0 \ \ \text{for each } i = 0, 1, 2.$$

So, by Theorem 14.1, we get, for all $f \in C_{\mathcal{F}}[0, 1]$, that

$$st_{C_1} - \lim_n D^* \left(B_n^{\mathcal{F}}(f), f \right) = 0.$$

However, since the sequence (u_n) given by (14.6) is non-convergent (in the usual sense), the sequence $\left(B_n^{\mathcal{F}}(f) \right)$ is not fuzzy convergent to f.

14.2 Statistical Fuzzy Rates

This section is devoted to studying the rates of A-statistical fuzzy convergence in Theorem 14.1. Before starting, we recall that various ways of defining rates of convergence in the A-statistical sense have been introduced in [62] as follows:

Let $A = [a_{jn}]$ be a non-negative regular summability matrix and let (p_n) be a positive non-increasing sequence of real numbers. Then

(a) A sequence $x = (x_n)$ is A-statistically convergent to the number L with the rate of $o(p_n)$ if for every $\varepsilon > 0$,

$$\lim_j \frac{1}{p_j} \sum_{n:|x_n-L|\geq\varepsilon} a_{jn} = 0.$$

In this case we write $x_n - L = st_A - o(p_n)$ as $n \to \infty$.

(b) If for every $\varepsilon > 0$,

$$\sup_{j} \frac{1}{p_j} \sum_{n:|x_n|\geq\varepsilon} a_{jn} < \infty,$$

then (x_n) is A−statistically bounded with the rate of $O(p_n)$ and it is denoted by $x_n = st_A - O(p_n)$ as $n \to \infty$.

(c) (x_n) is A-statistically convergent to L with the rate of $o_m(p_n)$, denoted by $x_n - L = st_A - o_m(p_n)$ as $n \to \infty$, if for every $\varepsilon > 0$,

$$\lim_{j} \sum_{n:|x_n-L|\geq\varepsilon p_n} a_{jn} = 0.$$

(d) (x_n) is A-statistically bounded with the rate of $O_m(p_n)$ provided that there is a positive number M satisfying

$$\lim_{j} \sum_{n:|x_n|\geq M p_n} a_{jn} = 0,$$

which is denoted by $x_n = st_A - O_m(p_n)$ as $n \to \infty$.

Unfortunately, there is no single definition that can become the standard for the comparison of the rates of summability transforms. The situation becomes even more uncharted when one considers the rates of A-statistical convergence. Observe that, in definitions (a) and (b), the "rate" is more controlled by the entries of the summability method rather than the terms of the sequence (x_n). For example, when one takes the identity matrix I, if (p_n) is any non-increasing sequence satisfying $1/p_n \leq M$ for some $M > 0$ and for each $n \in \mathbb{N}$, then $x_n - L = st_A - o(p_n)$ as $n \to \infty$ for any convergent sequence $(x_n - L)$ regardless of how slowly it goes to zero. To avoid such an unfortunate situation one may borrow the concept of convergence in measure from measure theory to define the rate of convergence as in definitions (c) and (d). So, we use the notations o_m and O_m, respectively.

Notice that, for the convergence of fuzzy number valued sequences or fuzzy number valued function sequences, we have to use the metrics D and D^* instead of the absolute value metric in all definitions mentioned above. In this case, for instance, we use the notation: $D(\mu_n, \mu) = st_A - o(p_n)$ as $n \to \infty$, where (μ_n) is a fuzzy number valued sequence, μ is a fuzzy number, and (p_n) is a positive non-increasing sequence of real numbers.

Let $f : [a, b] \to \mathbb{R}_{\mathcal{F}}$. Then the (first) fuzzy modulus of continuity of f, which is introduced by [82] (see also [6]), is defined by

$$\omega_1^{(\mathcal{F})}(f, \delta) := \sup_{x,y\in[a,b];\ |x-y|\leq\delta} D(f(x), f(y))$$

for any $0 < \delta \leq b - a$.

With this terminology, we get the next result.

Theorem 14.3. *Let $A = [a_{jn}]$ be a non-negative regular summability matrix and let (L_n) be a sequence of fuzzy positive linear operators from $C_{\mathcal{F}}[a, b]$ into itself. Assume that there exists a corresponding sequence $(\tilde{L}_n)$ of positive linear operators from $C[a, b]$ into itself with the property (14.1). Suppose that (a_n) and (b_n) are positive non-increasing sequences and also that the operators $\tilde{L}_n$ satisfy the following conditions:*

(i) $\left\| \tilde{L}_n(e_0) - e_0 \right\| = st_A - o(a_n)$ *as* $n \to \infty$,

(ii) $\omega_1^{(\mathcal{F})}(f, \mu_n) = st_A - o(b_n)$ *as* $n \to \infty$, *where* $\mu_n = \sqrt{\left\| \tilde{L}_n(\varphi) \right\|}$ *and*
$\varphi(y) = (y - x)^2$ *for each* $x \in [a, b]$.

Then, for all $f \in C_{\mathcal{F}}[a, b]$, *we get*

$$D^*(L_n(f), f) = st_A - o(c_n) \text{ as } n \to \infty,$$

where $c_n := \max\{a_n, b_n\}$ *for each* $n \in \mathbb{N}$. *Furthermore, similar results hold when little "o" is replaced by big "O".*

Proof. By Theorem 3 of [6], one can obtain, for each $n \in \mathbb{N}$ and $f \in C_{\mathcal{F}}[a, b]$, that

$$D^*(L_n(f), f) \leq M \left\| \tilde{L}_n(e_0) - e_0 \right\| + \left\| \tilde{L}_n(e_0) + e_0 \right\| \omega_1^{(\mathcal{F})}(f, \mu_n),$$

where $M := D^*\left(f, \chi_{\{0\}}\right)$ and $\chi_{\{0\}}$ denotes the neutral element for $\oplus$. Then we can write

$$D^*(L_n(f), f) \leq M \left\| \tilde{L}_n(e_0) - e_0 \right\| + \left\| \tilde{L}_n(e_0) - e_0 \right\| \omega_1^{(\mathcal{F})}(f, \mu_n) + 2\omega_1^{(\mathcal{F})}(f, \mu_n).$$
$$(14.7)$$

Now, for a given $\varepsilon > 0$, define the following sets:

$$V := \left\{ n \in \mathbb{N} : D^*(L_n(f), f) \geq \varepsilon \right\},$$
$$V_0 := \left\{ n \in \mathbb{N} : \left\| \tilde{L}_n(e_0) - e_0 \right\| \geq \frac{\varepsilon}{3M} \right\},$$
$$V_1 := \left\{ n \in \mathbb{N} : \left\| \tilde{L}_n(e_0) - e_0 \right\| \omega_1^{(\mathcal{F})}(f, \mu_n) \geq \frac{\varepsilon}{3} \right\},$$
$$V_2 := \left\{ n \in \mathbb{N} : \omega_1^{(\mathcal{F})}(f, \mu_n) \geq \frac{\varepsilon}{6} \right\}.$$

Then, by (14.7), we see that $V \subseteq V_0 \cup V_1 \cup V_2$. Also, considering

$$V_1' := \left\{ n \in \mathbb{N} : \left\| \tilde{L}_n(e_0) - e_0 \right\| \geq \sqrt{\frac{\varepsilon}{3}} \right\},$$
$$V_1'' := \left\{ n \in \mathbb{N} : \omega_1^{(\mathcal{F})}(f, \mu_n) \geq \sqrt{\frac{\varepsilon}{3}} \right\},$$

we immediately observe that $V_1 \subseteq V_1' \cup V_1''$, which implies $V \subseteq V_0 \cup V_1' \cup V_1'' \cup V_2$. Then, we have

$$\frac{1}{c_j} \sum_{n \in V} a_{jn} \le \frac{1}{c_j} \sum_{n \in V_0} a_{jn} + \frac{1}{c_j} \sum_{n \in V_1'} a_{jn} + \frac{1}{c_j} \sum_{n \in V_1''} a_{jn} + \frac{1}{c_j} \sum_{n \in V_2} a_{jn}. \quad (14.8)$$

Since $c_j = \max\{a_j, b_j\}$, we get from (14.8) that

$$\frac{1}{c_n} \sum_{n \in V} a_{jn} \le \frac{1}{a_j} \sum_{n \in V_0} a_{jn} + \frac{1}{a_j} \sum_{n \in V_1'} a_{jn} + \frac{1}{b_j} \sum_{n \in V_1''} a_{jn} + \frac{1}{b_j} \sum_{n \in V_2} a_{jn}. \quad (14.9)$$

So, letting $j \to \infty$ in (14.9) and using the hypotheses (i) and (ii), we have

$$\lim_j \frac{1}{c_j} \sum_{n \in V} a_{jn} = 0,$$

which finishes the proof.   ∎

By a similar way as in the proof of Theorem 14.3, one can obtain the next result at once.

Theorem 14.4. *Let $A = [a_{jn}]$ be a non-negative regular summability matrix and let (L_n) be a sequence of fuzzy positive linear operators from $C_{\mathcal{F}}[a, b]$ into itself. Assume that there exists a corresponding sequence $(\tilde{L}_n)$ of positive linear operators from $C[a, b]$ into itself with the property (14.1). Suppose that (a_n) and (b_n) are positive non-increasing sequences and also that the operators $\tilde{L}_n$ satisfy the following conditions:*

(i) $\left\| \tilde{L}_n(e_0) - e_0 \right\| = st_A - o_m(a_n)$ *as $n \to \infty$,*

(ii) $\omega_1^{(\mathcal{F})}(f, \mu_n) = st_A - o_m(b_n)$ *as $n \to \infty$.*

Then, for all $f \in C_{\mathcal{F}}[a, b]$, we get

$$D^* (L_n(f), f) = st_A - o(d_n) \text{ as } n \to \infty,$$

where $d_n := \max\{a_n, b_n, a_n b_n\}$ for each $n \in \mathbb{N}$. Furthermore, similar results hold when little "o_m" is replaced by big "O_m".

15

Statistical Fuzzy Trigonometric Korovkin-Type Approximation Theory

In this chapter, we consider non-negative regular summability matrix transformations in the approximation by fuzzy positive linear operators, where the test functions are trigonometric. So, we mainly obtain a trigonometric fuzzy Korovkin theorem by means of A-statistical convergence. We also compute the rates of A-statistical convergence of a sequence of fuzzy positive linear operators in the trigonometric environment. This chapter relies on [59].

15.1 Statistical Fuzzy Trigonometric Korovkin Theory

In this section we obtain a fuzzy trigonometric Korovkin-type approximation theorem by means of A-statistical convergence. Also, we give an example of fuzzy positive linear operators by using fuzzy Fejer operators, which express the importance of the statistical approximation.

By $C_{2\pi}^{(\mathcal{F})}(\mathbb{R})$ we mean the space of all fuzzy continuous and 2π-periodic functions on $\mathbb{R}$. Also the space of all real valued continuous and 2π-periodic functions is denoted by $C_{2\pi}(\mathbb{R})$. Assume that $f : [a, b] \to \mathbb{R}_{\mathcal{F}}$ is a fuzzy number valued function. Then, f is said to be fuzzy-Riemann integrable (or, FR-integrable) to $I \in \mathbb{R}_{\mathcal{F}}$ if, for given $\varepsilon > 0$, there exists a $\delta > 0$ such that, for any partition $P = \{[u, v]; \xi\}$ of $[a, b]$ with the norms $\Delta(P) < \delta$,

G.A. Anastassiou and O. Duman: Towards Intelligent Modeling, ISRL 14, pp. 189–197.
springerlink.com   © Springer-Verlag Berlin Heidelberg 2011

we get

$$D\left(\bigoplus_P (v-u) \odot f(\xi), I\right) < \varepsilon.$$

In this case, we write

$$I := (FR)\int_a^b f(x)dx.$$

By Corollary 13.2 of [82, p. 644], we deduce that if $f \in C_{\mathcal{F}}[a,b]$ (fuzzy continuous on $[a,b]$), then f is FR-integrable on $[a,b]$. Now let $L : C_{2\pi}^{(\mathcal{F})}(\mathbb{R}) \to C_{2\pi}^{(\mathcal{F})}(\mathbb{R})$ be an operator. Then L is said to be fuzzy linear if, for every $\lambda_1, \lambda_2 \in \mathbb{R}$, $f_1, f_2 \in C_{\mathcal{F}}(\mathbb{R})$, and $x \in \mathbb{R}$,

$$L\left(\lambda_1 \odot f_1 \oplus \lambda_2 \odot f_2; x\right) = \lambda_1 \odot L(f_1; x) \oplus \lambda_2 \odot L(f_2; x)$$

holds. Also L is called fuzzy positive linear operator if it is fuzzy linear and, the condition $L(f; x) \preceq L(g; x)$ is satisfied for any $f, g \in C_{\mathcal{F}}(\mathbb{R})$ and all $x \in \mathbb{R}$ with $f(x) \preceq g(x)$.

Throughout this section we use the test functions f_i ($i = 0, 1, 2$) defined by

$$f_0(x) = 1, \ f_1(x) = \cos x, \ f_2(x) = \sin x.$$

Then, we obtain the next result.

Theorem 15.1. *Let $A = [a_{jn}]$ be a non-negative regular summability matrix and let (L_n) be a sequence of fuzzy positive linear operators defined on $C_{2\pi}^{(\mathcal{F})}(\mathbb{R})$. Assume that there exists a corresponding sequence $(\tilde{L}_n)$ of positive linear operators defined on $C_{2\pi}(\mathbb{R})$ with the property*

$$\{L_n(f; x)\}_{\pm}^{(r)} = \tilde{L}_n\left(f_{\pm}^{(r)}; x\right) \tag{15.1}$$

for all $x \in [a,b]$, $r \in [0,1]$, $n \in \mathbb{N}$ and $f \in C_{2\pi}^{(\mathcal{F})}(\mathbb{R})$. Assume further that

$$st_A - \lim_n \left\|\tilde{L}_n(f_i) - f_i\right\| = 0 \ \text{ for each } i = 0, 1, 2, \tag{15.2}$$

the symbol $\|g\|$ denotes the usual sup-norm of $g \in C_{2\pi}(\mathbb{R})$. Then, for all $f \in C_{2\pi}^{(\mathcal{F})}(\mathbb{R})$, we get

$$st_A - \lim_n D^*\left(L_n(f), f\right) = 0.$$

Proof. Assume that I is a closed bounded interval with length 2π of $\mathbb{R}$. Now let $f \in C_{2\pi}^{(\mathcal{F})}(\mathbb{R})$, $x \in I$ and $r \in [0,1]$. Taking $[f(x)]^{(r)} = \left[f_-^{(r)}(x), \ f_+^{(r)}(x)\right]$ we get $f_{\pm}^{(r)} \in C_{2\pi}(\mathbb{R})$. Hence, for every $\varepsilon > 0$, there exists a $\delta > 0$ such that

$$\left|f_{\pm}^{(r)}(y) - f_{\pm}^{(r)}(x)\right| < \varepsilon \tag{15.3}$$

for all y satisfying $|y - x| < \delta$. On the other hand, by the boundedness of $f_{\pm}^{(r)}$, the following inequality

$$\left| f_{\pm}^{(r)}(y) - f_{\pm}^{(r)}(x) \right| \leq 2 \left\| f_{\pm}^{(r)} \right\| \tag{15.4}$$

holds for all $y \in \mathbb{R}$. Now consider the subinterval $(x - \delta, 2\pi + x - \delta]$ with length 2π. Then, by (15.3) and (15.4), it is easy to check that

$$\left| f_{\pm}^{(r)}(y) - f_{\pm}^{(r)}(x) \right| \leq \varepsilon + 2M_{\pm}^{(r)} \frac{\varphi(y)}{\sin^2 \frac{\delta}{2}} \tag{15.5}$$

holds for all $y \in (x - \delta, 2\pi + x - \delta]$, where $\varphi(y) := \sin^2\left(\frac{y-x}{2}\right)$ and $M_{\pm}^{(r)} := \left\| f_{\pm}^{(r)} \right\|$. Observe that inequality (15.5) also holds for all $y \in \mathbb{R}$ because of the periodicity of $f_{\pm}^{(r)}$ (see, for instance, [93]). Now using the linearity and the positivity of the operators $\tilde{L}_n$ and considering inequality (15.5), we can write, for each $n \in \mathbb{N}$, that

$$\left| \tilde{L}_n\left(f_{\pm}^{(r)}; x\right) - f_{\pm}^{(r)}(x) \right| \leq \tilde{L}_n\left(\left| f_{\pm}^{(r)}(y) - f_{\pm}^{(r)}(x) \right|; x \right)$$
$$+ M_{\pm}^{(r)} \left| \tilde{L}_n\left(f_0; x\right) - f_0(x) \right|$$
$$\leq \varepsilon + \left(\varepsilon + M_{\pm}^{(r)} \right) \left| \tilde{L}_n\left(f_0; x\right) - f_0(x) \right|$$
$$+ \frac{2M_{\pm}^{(r)}}{\sin^2 \frac{\delta}{2}} \left| \tilde{L}_n\left(\varphi; x\right) \right|.$$

Hence, we see that

$$\left| \tilde{L}_n\left(f_{\pm}^{(r)}; x\right) - f_{\pm}^{(r)}(x) \right| \leq \varepsilon + \left(\varepsilon + M_{\pm}^{(r)} + \frac{2M_{\pm}^{(r)}}{\sin^2 \frac{\delta}{2}} \right) \left| \tilde{L}_n\left(f_0; x\right) - f_0(x) \right|$$
$$+ \frac{2M_{\pm}^{(r)}}{\sin^2 \frac{\delta}{2}} \left| \tilde{L}_n\left(f_1; x\right) - f_1(x) \right|$$
$$+ \frac{2M_{\pm}^{(r)}}{\sin^2 \frac{\delta}{2}} \left| \tilde{L}_n\left(f_2; x\right) - f_2(x) \right|.$$

Letting $K_{\pm}^{(r)}(\varepsilon) := \varepsilon + M_{\pm}^{(r)} + \frac{2M_{\pm}^{(r)}}{\sin^2 \frac{\delta}{2}}$ and taking supremum over $x \in \mathbb{R}$, we easily observe that

$$\left\| \tilde{L}_n\left(f_{\pm}^{(r)}\right) - f_{\pm}^{(r)} \right\| \leq \varepsilon + K_{\pm}^{(r)}(\varepsilon)\{ \left\| \tilde{L}_n\left(f_0\right) - f_0 \right\|$$
$$+ \left\| \tilde{L}_n\left(f_1\right) - f_1 \right\| + \left\| \tilde{L}_n\left(f_2\right) - f_2 \right\| \}. \tag{15.6}$$

Now, from (15.6), we derive that

$$
\begin{aligned}
D^{*}\left(L_{n}(f), f\right) &= \sup_{x \in \mathbb{R}} D\left(L_{n}(f ; x), f(x)\right) \\
&= \sup_{x \in \mathbb{R}} \sup_{r \in [0,1]} \max \left\{\left|\tilde{L}_{n}\left(f_{-}^{(r)} ; x\right)-f_{-}^{(r)}(x)\right|,\left|\tilde{L}_{n}\left(f_{+}^{(r)} ; x\right)-f_{+}^{(r)}(x)\right|\right\} \\
&= \sup_{r \in [0,1]} \max \left\{\left\|\tilde{L}_{n}\left(f_{-}^{(r)}\right)-f_{-}^{(r)}\right\|,\left\|\tilde{L}_{n}\left(f_{+}^{(r)}\right)-f_{+}^{(r)}\right\|\right\} .
\end{aligned}
$$

Therefore, combining the above equality with (15.6), we get

$$
\begin{aligned}
D^{*}\left(L_{n}(f), f\right) \leq \varepsilon+K(\varepsilon)\{&\left\|\tilde{L}_{n}\left(f_{0}\right)-f_{0}\right\|+\left\|\tilde{L}_{n}\left(f_{1}\right)-f_{1}\right\| \\
&+\left\|\tilde{L}_{n}\left(f_{2}\right)-f_{2}\right\|\},
\end{aligned} \tag{15.7}
$$

where $K(\varepsilon):=\sup_{r \in [0,1]} \max \left\{K_{-}^{(r)}(\varepsilon), K_{+}^{(r)}(\varepsilon)\right\}$. Now, for a given $\varepsilon^{\prime}>0$, chose $\varepsilon>0$ such that $0<\varepsilon<\varepsilon^{\prime}$, and define the following sets

$$
\begin{aligned}
U &:=\left\{n \in \mathbb{N}: D^{*}\left(L_{n}(f), f\right) \geq \varepsilon^{\prime}\right\}, \\
U_{0} &:=\left\{n \in \mathbb{N}:\left\|\tilde{L}_{n}\left(f_{0}\right)-f_{0}\right\| \geq \frac{\varepsilon^{\prime}-\varepsilon}{3 K(\varepsilon)}\right\}, \\
U_{1} &:=\left\{n \in \mathbb{N}:\left\|\tilde{L}_{n}\left(f_{1}\right)-f_{1}\right\| \geq \frac{\varepsilon^{\prime}-\varepsilon}{3 K(\varepsilon)}\right\}, \\
U_{2} &:=\left\{n \in \mathbb{N}:\left\|\tilde{L}_{n}\left(f_{2}\right)-f_{2}\right\| \geq \frac{\varepsilon^{\prime}-\varepsilon}{3 K(\varepsilon)}\right\}.
\end{aligned}
$$

Then inequality (15.4) implies

$$
U \subseteq U_{0} \cup U_{1} \cup U_{2},
$$

which gives that, for each $j \in \mathbb{N}$,

$$
\sum_{n \in U} a_{jn} \leq \sum_{n \in U_{0}} a_{jn}+\sum_{n \in U_{1}} a_{jn}+\sum_{n \in U_{2}} a_{jn}.
$$

If we take limit as $j \rightarrow \infty$ on the both sides of inequality (15.6) and use the hypothesis (15.2), we immediately obtain that

$$
\lim_{j} \sum_{n \in U} a_{jn}=0,
$$

which completes the proof. ∎

Concluding Remarks

1. If we replace the matrix A in Theorem 15.1 by the Cesáro matrix C_{1}, we immediately have the next statistical fuzzy Korovkin result in the trigonometric case.

Corollary 15.2. *Let (L_n) be a sequence of fuzzy positive linear operators defined on $C_{2\pi}^{(\mathcal{F})}(\mathbb{R})$, and let $(\tilde{L}_n)$ be a corresponding sequence of positive linear operators defined on $C_{2\pi}(\mathbb{R})$ with the property (15.1). Assume that*

$$st - \lim_n \left\| \tilde{L}_n(f_i) - f_i \right\| = 0 \ \ \text{for each } i = 0, 1, 2.$$

Then, for all $f \in C_{2\pi}^{(\mathcal{F})}(\mathbb{R})$, we get

$$st - \lim_n D^* \left(L_n(f), f \right) = 0.$$

2. Replacing the matrix A by the identity matrix, one can get the classical fuzzy Korovkin result which was introduced by Anastassiou and Gal [36].

Corollary 15.3 (see [36]). *Let (L_n) be a sequence of fuzzy positive linear operators defined on $C_{2\pi}^{(\mathcal{F})}(\mathbb{R})$, and let $(\tilde{L}_n)$ be a corresponding sequence of positive linear operators defined on $C_{2\pi}(\mathbb{R})$ with the property (15.1). Assume that the sequence $\left(\tilde{L}_n(f_i) \right)$ is uniformly convergent to f_i on the whole real line (in the ordinary sense). Then, for all $f \in C_{2\pi}^{(\mathcal{F})}(\mathbb{R})$, the sequence $(L_n(f))$ is uniformly convergent to f on the whole real line (in the fuzzy sense).*

3. Now the following application shows that the A-statistical fuzzy Korovkin-type approximation theorem in the trigonometric case (Theorem 15.1) is a non-trivial generalization of its classical case (Corollary 15.3) given by Anastassiou and Gal [36].

Let $A = [a_{jn}]$ be any non-negative regular summability matrix. Assume that K is any subset of $\mathbb{N}$ satisfying $\delta_A(K) = 0$. Then define a sequence (u_n) by:

$$u_n = \begin{cases} \sqrt{n}, & \text{if } n \in K \\ 0, & \text{if } n \in \mathbb{N} \backslash K. \end{cases} \tag{15.8}$$

In this case, we see that (u_n) is non-convergent (in the ordinary sense). However, since for every $\varepsilon > 0$

$$\lim_j \sum_{n:|u_n| \geq \varepsilon} a_{jn} = \lim_j \sum_{n \in K} a_{jn} = \delta_A(K) = 0,$$

we have

$$st_A - \lim_n u_n = 0, \tag{15.9}$$

although the sequence (u_n) is unbounded from above. Now consider the fuzzy Fejer operators F_n as follows:

$$F_n(f; x) := \frac{1}{n\pi} \odot \left\{ (FR) \int_{-\pi}^{\pi} f(y) \odot \frac{\sin^2\left(\frac{n}{2}(y - x)\right)}{2 \sin^2\left[\left(\frac{y-x}{2}\right)\right]} dy \right\}, \tag{15.10}$$

where $n \in \mathbb{N}$, $f \in C_{2\pi}^{(\mathcal{F})}(\mathbb{R})$ and $x \in \mathbb{R}$. Then observe that the operators F_n are fuzzy positive linear. Also, the corresponding real Fejer operators have the following form:

$$\{F_n(f;x)\}_{\pm}^{(r)} = \tilde{F}_n\left(f_{\pm}^{(r)};x\right) := \frac{1}{n\pi} \int\limits_{-\pi}^{\pi} f_{\pm}(y) \frac{\sin^2\left(\frac{n}{2}(y-x)\right)}{2\sin^2\left[\left(\frac{y-x}{2}\right)\right]} dy$$

where $f_{\pm}^{(r)} \in C_{2\pi}(\mathbb{R})$ and $r \in [0,1]$. Then, we get (see [93])

$$\tilde{F}_n\left(f_0;x\right) = 1,$$

$$\tilde{F}_n\left(f_1;x\right) = \frac{n-1}{n}\cos x,$$

$$\tilde{F}_n\left(f_2;x\right) = \frac{n-1}{n}\sin x.$$

Now using the sequence (u_n) given by (15.8) we define the following fuzzy positive linear operators on the space $C_{2\pi}^{(\mathcal{F})}(\mathbb{R})$:

$$T_n(f;x) := (1+u_n) \odot F_n(f;x), \qquad (15.11)$$

where $n \in \mathbb{N}$, $f \in C_{2\pi}^{(\mathcal{F})}(\mathbb{R})$ and $x \in \mathbb{R}$. So, the corresponding real positive linear operators are given by

$$\tilde{T}_n\left(f_{\pm}^{(r)};x\right) := \frac{1+u_n}{n\pi} \int\limits_{-\pi}^{\pi} f_{\pm}(y) \frac{\sin^2\left(\frac{n}{2}(y-x)\right)}{2\sin^2\left[\left(\frac{y-x}{2}\right)\right]} dy,$$

where $f_{\pm}^{(r)} \in C_{2\pi}(\mathbb{R})$. Then we get, for all $n \in \mathbb{N}$ and $x \in \mathbb{R}$, that

$$\left\|\tilde{T}_n\left(f_0\right) - f_0\right\| = u_n,$$

$$\left\|\tilde{T}_n\left(f_1\right) - f_1\right\| \leq u_n + \frac{1+u_n}{n},$$

$$\left\|\tilde{T}_n\left(f_2\right) - f_2\right\| \leq u_n + \frac{1+u_n}{n}.$$

It follows from (15.9) that

$$st_A - \lim_n \left\|\tilde{T}_n\left(f_0\right) - f_0\right\| = 0. \qquad (15.12)$$

Also, by the definition of (u_n) we get

$$\lim_n \frac{1+u_n}{n} = 0,$$

which yields, for any non-negative regular matrix $A = [a_{jn}]$, that

$$st_A - \lim_n \frac{1+u_n}{n} = 0. \qquad (15.13)$$

Now by (15.9) and (15.13) we easily observe that, for every $\varepsilon > 0$,

$$\lim_j \sum_{n:\left\|\tilde{T}_n(f_1)-f_1\right\|\geq\varepsilon} a_{jn} \leq \lim_j \sum_{n:|u_n|\geq\frac{\varepsilon}{2}} a_{jn} + \lim_j \sum_{n:\left|\frac{1+u_n}{n}\right|\geq\frac{\varepsilon}{2}} a_{jn} = 0.$$

So we have

$$st_A - \lim_n \left\|\tilde{T}_n(f_1) - f_1\right\| = 0. \tag{15.14}$$

By a similar idea, one can derive that

$$st_A - \lim_n \left\|\tilde{T}_n(f_2) - f_2\right\| = 0. \tag{15.15}$$

Now, with the help of (15.12), (15.14), (15.15), all hypotheses of Theorem 2.1 hold. Then, we deduce, for all $f \in C_{2\pi}^{(\mathcal{F})}(\mathbb{R})$, that

$$st_A - \lim_n D^*(T_n(f), f) = 0.$$

However, since the sequence (u_n) is non-convergent and also unbounded from above, the sequence $(T_n(f))$ is not fuzzy convergent to f. Hence, Corollary 15.3 does not work for the operators T_n defined by (15.11).

15.2 Statistical Fuzzy Rates in Trigonometric Case

Let $A = [a_{jn}]$ be a non-negative regular summability matrix and let (p_n) be a positive non-increasing sequence of real numbers. We know from [62] that a sequence (x_n) is A-statistically convergent to the number L with the rate of $o(p_n)$ if for every $\varepsilon > 0$, $\lim_j \frac{1}{p_j}\sum_{n:|x_n-L|\geq\varepsilon} a_{jn} = 0$, which is denoted by $x_n - L = st_A - o(p_n)$ as $n \to \infty$. If, for every $\varepsilon > 0$, $\sup_j \frac{1}{p_j}\sum_{n:|x_n|\geq\varepsilon} a_{jn} < \infty$, then (x_n) is A-statistically bounded with the rate of $O(p_n)$, which is denoted by $x_n = st_A - O(p_n)$ as $n \to \infty$. If, for every $\varepsilon > 0$, $\lim_j \sum_{n:|x_n-L|\geq\varepsilon p_n} a_{jn} = 0$, then (x_n) is A-statistically convergent to L with the rate of $o_m(p_n)$, which is denoted by $x_n - L = st_A - o_m(p_n)$ as $n \to \infty$. Finally, if there is a positive number M satisfying $\lim_j \sum_{n:|x_n|\geq Mp_n} a_{jn} = 0$, then we say that (x_n) is A-statistically bounded with the rate of $O_m(p_n)$, which is denoted by $x_n - L = st_A - O_m(p_n)$ as $n \to \infty$. We should also note that, for the convergence of fuzzy number valued sequences or fuzzy number valued function sequences, we have to use the metrics D and D^* instead of the absolute value metric in all definitions mentioned above.

Let $f \in C_{2\pi}^{(\mathcal{F})}(\mathbb{R})$. Then, the quantity $\omega_1^{(\mathcal{F})}(f, \delta)$, $\delta > 0$, denotes the (first) fuzzy modulus of continuity of f, which was introduced by [82] (see also

[6, 36]) as follows:

$$\omega_1^{(\mathcal{F})}(f, \delta) := \sup_{x, y \in \mathbb{R};\ |x - y| \leq \delta} D\left(f(x), f(y)\right)$$

for any $\delta > 0$.

So, we get the next result.

Theorem 15.4. *Let $A = [a_{jn}]$ be a non-negative regular summability matrix and let (L_n) be a sequence of fuzzy positive linear operators defined on $C_{2\pi}^{(\mathcal{F})}(\mathbb{R})$. Assume that there exists a corresponding sequence $(\tilde{L}_n)$ of positive linear operators on $C_{2\pi}(\mathbb{R})$ with the property (15.1). Suppose that (a_n) and (b_n) are positive non-increasing sequences and also that the operators $\tilde{L}_n$ satisfy the following conditions:*

(i) $\left\| \tilde{L}_n(f_0) - f_0 \right\| = st_A - o(a_n)$ *as* $n \to \infty$,

(ii) $\omega_1^{(\mathcal{F})}(f, \mu_n) = st_A - o(b_n)$ *as* $n \to \infty$, *where* $\mu_n = \sqrt{\left\| \tilde{L}_n(\varphi) \right\|}$ *and* $\varphi(y) = \sin^2\left(\frac{y-x}{2}\right)$ *for each* $x \in \mathbb{R}$.

Then, for all $f \in C_{2\pi}^{(\mathcal{F})}(\mathbb{R})$, we get

$$D^*\left(L_n(f), f\right) = st_A - o(c_n) \text{ as } n \to \infty,$$

where $c_n := \max\{a_n, b_n\}$ for each $n \in \mathbb{N}$. Furthermore, similar results hold when little "o" is replaced by big "O".

Proof. Let $f \in C_{2\pi}^{(\mathcal{F})}(\mathbb{R})$. Then, using the property (15.1) and applying Theorem 4 of [36], we immediately observe, for each $n \in \mathbb{N}$, that

$$D^*\left(L_n(f), f\right) \leq M \left\| \tilde{L}_n(f_0) - f_0 \right\| + \left\| \tilde{L}_n(f_0) + f_0 \right\| \omega_1^{(\mathcal{F})}(f, \mu_n),$$

where $M := D^*\left(f, \chi_{\{0\}}\right)$ and $\chi_{\{0\}}$ denotes the neutral element for $\oplus$. The last inequality implies that

$$D^*\left(L_n(f), f\right) \leq M \left\| \tilde{L}_n(f_0) - f_0 \right\| + \left\| \tilde{L}_n(f_0) - f_0 \right\| \omega_1^{(\mathcal{F})}(f, \mu_n) + 2\omega_1^{(\mathcal{F})}(f, \mu_n) \tag{15.16}$$

holds for each $n \in \mathbb{N}$. Now, for a given $\varepsilon > 0$, define the following sets:

$$V := \left\{ n \in \mathbb{N} : D^*\left(L_n(f), f\right) \geq \varepsilon \right\},$$

$$V_0 := \left\{ n \in \mathbb{N} : \left\| \tilde{L}_n(f_0) - f_0 \right\| \geq \frac{\varepsilon}{3M} \right\},$$

$$V_1 := \left\{ n \in \mathbb{N} : \left\| \tilde{L}_n(f_0) - f_0 \right\| \geq \sqrt{\frac{\varepsilon}{3}} \right\},$$

$$V_2 := \left\{ n \in \mathbb{N} : \omega_1^{(\mathcal{F})}(f, \mu_n) \geq \sqrt{\frac{\varepsilon}{3}} \right\},$$

$$V_3 := \left\{ n \in \mathbb{N} : \omega_1^{(\mathcal{F})}(f, \mu_n) \geq \frac{\varepsilon}{6} \right\}.$$

Thus, inequality (15.16) gives that $V \subseteq V_0 \cup V_1 \cup V_2 \cup V_3$. Then we can write, for each $j \in \mathbb{N}$, that

$$\frac{1}{c_j} \sum_{n \in V} a_{jn} \leq \frac{1}{c_j} \sum_{n \in V_0} a_{jn} + \frac{1}{c_j} \sum_{n \in V_1'} a_{jn} + \frac{1}{c_j} \sum_{n \in V_1''} a_{jn} + \frac{1}{c_j} \sum_{n \in V_2} a_{jn}. \quad (15.17)$$

Also using the fact $c_j = \max\{a_j, b_j\}$, we get from (15.17) that

$$\frac{1}{c_j} \sum_{n \in V} a_{jn} \leq \frac{1}{a_j} \sum_{n \in V_0} a_{jn} + \frac{1}{a_j} \sum_{n \in V_1'} a_{jn} + \frac{1}{b_j} \sum_{n \in V_1''} a_{jn} + \frac{1}{b_j} \sum_{n \in V_2} a_{jn}. \quad (15.18)$$

Therefore, taking limit as $j \to \infty$ on the both sides of inequality (15.18) and using the hypotheses (i) and (ii), we obtain that

$$\lim_j \frac{1}{c_j} \sum_{n \in V} a_{jn} = 0,$$

which yields that

$$st_A - \lim_n D^* \left(L_n(f), f \right) = 0$$

for all $f \in C_{2\pi}^{(\mathcal{F})}(\mathbb{R})$.   ∎

One can also prove the following analog.

Theorem 15.5. *Let $A = [a_{jn}]$ be a non-negative regular summability matrix and let (L_n) be a sequence of fuzzy positive linear operators on $C_{2\pi}^{(\mathcal{F})}(\mathbb{R})$. Assume that there exists a corresponding sequence $(\tilde{L}_n)$ of positive linear operators on $C_{2\pi}(\mathbb{R})$ with the property (15.1). Suppose that (a_n) and (b_n) are positive non-increasing sequences and also that the operators $\tilde{L}_n$ satisfy the following conditions:*

(i) $\left\| \tilde{L}_n(f_0) - f_0 \right\| = st_A - o_m(a_n)$ as $n \to \infty$,

(ii) $\omega_1^{(\mathcal{F})}(f, \mu_n) = st_A - o_m(b_n)$ as $n \to \infty$, where μ_n is given as in Theorem 15.4.

Then, for all $f \in C_{2\pi}^{(\mathcal{F})}(\mathbb{R})$, we get

$$D^* \left(L_n(f), f \right) = st_A - o(d_n) \text{ as } n \to \infty,$$

where $d_n := \max\{a_n, b_n, a_n b_n\}$ for each $n \in \mathbb{N}$. Furthermore, similar results hold when little "o_m" is replaced by big "O_m".

16

High Order Statistical Fuzzy Korovkin-Type Approximation Theory

In this chapter, we obtain a statistical fuzzy Korovkin-type approximation result with high rate of convergence. Main tools used in this work are statistical convergence and higher order continuously differentiable functions in the fuzzy sense. An application is also given, which demonstrates that the statistical fuzzy approximation is stronger than the classical one. This chapter relies on [22].

16.1 High Order Statistical Fuzzy Korovkin Theory

As usual, by $C_{\mathcal{F}}[a,b]$ we denote the space of all fuzzy continuous functions on $[a,b]$. Assume that $f : [a,b] \to \mathbb{R}_{\mathcal{F}}$ is a fuzzy number valued function. Then f is called (fuzzy) differentiable at $x \in [a,b]$ if there exists a $f'(x) \in \mathbb{R}_{\mathcal{F}}$ such that the following limits

$$\lim_{h \to 0^+} \frac{f(x+h) - f(x)}{h}, \quad \lim_{h \to 0^+} \frac{f(x) - f(x-h)}{h}$$

exist and are equal to $f'(x)$. If f is differentiable at any point $x \in [a,b]$, then we call that f is (fuzzy) differentiable on $[a,b]$ with the derivative f' (see [107]). Similarly, we can define higher order fuzzy derivatives. Also by $C_{\mathcal{F}}^{(m)}[a,b]$ ($m \in \mathbb{N}$) we mean all fuzzy valued functions from $[a,b]$ into $\mathbb{R}_{\mathcal{F}}$ that are m-times continuously differentiable in the fuzzy sense. Using these definitions Kaleva [88] proved the following result.

G.A. Anastassiou and O. Duman: Towards Intelligent Modeling, ISRL 14, pp. 199–206.
springerlink.com   © Springer-Verlag Berlin Heidelberg 2011

Lemma A (see [88]). *Let* $f : [a, b] \subset R \to R_{\mathcal{F}}$ *be fuzzy differentiable, and let* $x \in [a, b]$, $0 \le r \le 1$. *Then, clearly*

$$[f(x)]^r = [(f(x))_-^{(r)}, (f(x))_+^{(r)}] \subset \mathbb{R}.$$

Then $(f(x))_\pm^{(r)}$ *are differentiable and*

$$[f'(x)]^r = \left[\left((f(x))_-^{(r)}\right)', \left((f(x))_+^{(r)}\right)' \right],$$

i.e.,

$$(f')_\pm^{(r)} = \left(f_\pm^{(r)}\right)' \quad \text{for any } r \in [0, 1].$$

Also, for higher order fuzzy derivatives, Anastassiou [5] obtained the similar result:

Lemma B (see [5]). *Let* $m \in N$ *and* $f \in C_{\mathcal{F}}^{(m)}[a, b]$. *Then, we have* $f_\pm^{(r)} \in C^m[a, b]$ *(for any* $r \in [0, 1]$*) and*

$$[f^{(i)}(x)]^r = \left[\left((f(x))_-^{(r)}\right)^{(i)}, \left((f(x))_+^{(r)}\right)^{(i)} \right]$$

for $i = 0, 1, ..., m$, *and, in particular, we have*

$$(f^{(i)})_\pm^{(r)} = \left(f_\pm^{(r)}\right)^{(i)} \quad \text{for any } r \in [0, 1] \text{ and } i = 0, 1, ..., m.$$

We also recall that the (first) fuzzy modulus of continuity of $f \in C_{\mathcal{F}}[a, b]$, which is introduced by [82] (see also [6]), is given by

$$\omega_1^{(\mathcal{F})}(f, \delta) := \sup_{x, y \in [a, b]; \, |x-y| \le \delta} D\left(f(x), f(y)\right)$$

for any $0 < \delta \le b - a$.

We first need the next result.

Theorem A (see [5]). *Let* (L_n) *be a sequence of fuzzy positive linear operators from* $C_{\mathcal{F}}^{(m)}[a, b]$ *into* $C_{\mathcal{F}}[a, b]$. *Assume that there exists a corresponding sequence* $(\tilde{L}_n)$ *of positive linear operators from* $C^m[a, b]$ *into* $C[a, b]$ *with the property* $\{L_n(f; x)\}_\pm^{(r)} = \tilde{L}_n\left(f_\pm^{(r)}; x\right)$ *for all* $x \in [a, b]$, $r \in [0, 1]$, $n \in \mathbb{N}$ *and* $f \in C_{\mathcal{F}}^{(m)}[a, b]$. *Then, the following inequality*

$$D^*\left(L_n(f), f\right) \le M_0 \left\|\tilde{L}_n(e_0) - e_0\right\| + \sum_{k=1}^m \frac{M_k \left\|\tilde{L}_n(\Psi^k)\right\|}{k!}$$

$$+ \frac{\left\|\tilde{L}_n(\Psi^{m+1})\right\|^{\frac{m}{m+1}}}{m!} \left\|\left(\tilde{L}_n(e_0)\right)^{\frac{1}{m+1}} + \frac{1}{m+1}\right\|$$

$$\times \omega_1^{(\mathcal{F})}\left(f^{(m)}, \left\|\tilde{L}_n(\Psi^{m+1})\right\|^{\frac{1}{m+1}}\right)$$

holds, where $\|g\|$ denotes the sup-norm of $g \in C[a,b]$, $e_0(y) := 1$, $\Psi(y) := |y - x|$ for each $x \in [a, b]$, $M_0 := D(f, \chi_{\{0\}})$, $M_k := D(f^{(k)}, \chi_{\{0\}})$, $(k = 1, 2, ..., m)$, and $\chi_{\{0\}}$ denotes the neutral element for $\oplus$.

Lemma 16.1. Let $A = [a_{jn}]$ be a non-negative regular summability matrix and (δ_n) be a sequence of positive real numbers. If $st_A - \lim_n \delta_n = 0$, then we have $st_A - \lim_n \omega_1^{(\mathcal{F})}(f, \delta_n) = 0$ for all $f \in C_{\mathcal{F}}[a, b]$.

Proof. Let $f \in C_{\mathcal{F}}[a, b]$ be fixed. Since $st_A - \lim_n \delta_n = 0$, we obtain, for any $\delta > 0$, that

$$\lim_j \sum_{n:\delta_n \geq \delta} a_{jn} = 0. \tag{16.1}$$

By the right-continuity of $\omega_1^{(\mathcal{F})}(f, \cdot)$ at zero, we can write that, for a given $\varepsilon > 0$, there exists a $\delta > 0$ such that $\omega_1^{(\mathcal{F})}(f, \alpha) < \varepsilon$ whenever $\alpha < \delta$, i.e., $\omega_1^{(\mathcal{F})}(f, \alpha) \geq \varepsilon$ gives that $\alpha \geq \delta$. Now replacing α by δ_n, for every $\varepsilon > 0$, we observe that

$$\{n : \omega_1^{(\mathcal{F})}(f, \delta_n) \geq \varepsilon\} \subseteq \{n : \delta_n \geq \delta\}. \tag{16.2}$$

So, it follows from (16.2) that, for each $j \in \mathbb{N}$,

$$\sum_{n:\omega_1^{(\mathcal{F})}(f,\delta_n) \geq \varepsilon} a_{jn} \leq \sum_{n:\delta_n \geq \delta} a_{jn}. \tag{16.3}$$

Then, letting $j \to \infty$ on the both sides of inequality (16.3) and using (16.1) we immediately see, for every $\varepsilon > 0$, that

$$\lim_j \sum_{n:\omega_1^{(\mathcal{F})}(f,\delta_n) \geq \varepsilon} a_{jn} = 0$$

which implies that $st_A - \lim_n \omega_1^{(\mathcal{F})}(f, \delta_n) = 0$. So, the proof is done. ∎

Now we are ready to present the main result of this section.

Theorem 16.2. Let $A = [a_{jn}]$ be a non-negative regular summability matrix and let (L_n) be a sequence of fuzzy positive linear operators from $C_{\mathcal{F}}^{(m)}[a, b]$ into $C_{\mathcal{F}}[a, b]$. Assume that there exists a corresponding sequence $(\tilde{L}_n)$ of positive linear operators from $C^m[a, b]$ into $C[a, b]$ with the property

$$\{L_n(f; x)\}_{\pm}^{(r)} = \tilde{L}_n \left(f_{\pm}^{(r)}; x \right) \tag{16.4}$$

for all $x \in [a, b]$, $r \in [0, 1]$, $n \in \mathbb{N}$ and $f \in C_{\mathcal{F}}^{(m)}[a, b]$. Assume further that the following statements hold:

$$st_A - \lim_n \left\| \tilde{L}_n(e_0) - e_0 \right\| = 0 \quad \text{with } e_0(y) := 1 \tag{16.5}$$

and

$$st_A - \lim_n \left\| \tilde{L}_n(\Psi^{m+1}) \right\| = 0 \quad \text{with } \Psi(y) := |y - x| \text{ for each } x \in [a, b].$$
(16.6)

Then, for all $f \in C_{\mathcal{F}}^{(m)}[a, b]$, we get

$$st_A - \lim_n D^* (L_n(f), f) = 0.$$

Proof. Let $f \in C_{\mathcal{F}}^{(m)}[a, b]$. By Theorem A, we see, for all $n \in \mathbb{N}$, that

$$D^* (L_n(f), f) \leq M_0 \left\| \tilde{L}_n(e_0) - e_0 \right\| + \sum_{k=1}^{m} \frac{M_k \left\| \tilde{L}_n(\Psi^k) \right\|}{k!}$$
$$+ \frac{\left\| \tilde{L}_n(\Psi^{m+1}) \right\|^{\frac{m}{m+1}}}{m!} \left\| \left(\tilde{L}_n(e_0) \right)^{\frac{1}{m+1}} + \frac{1}{m+1} \right\| \quad (16.7)$$
$$\times \omega_1^{(\mathcal{F})} \left(f^{(m)}, \left\| \tilde{L}_n(\Psi^{m+1}) \right\|^{\frac{1}{m+1}} \right),$$

where $M_0 = D(f, \chi_{\{0\}})$ and $M_k = D(f^{(k)}, \chi_{\{0\}})$, $(k = 1, 2, ..., m)$, and $\chi_{\{0\}}$ denotes the neutral element for $\oplus$. Now using the Hölder inequality on the term $\left\| \tilde{L}_n(\Psi^k) \right\|$ with $p = \dfrac{m+1}{m+1-k}$ and $q = \dfrac{m+1}{k}$, where $\dfrac{1}{p} + \dfrac{1}{q} = 1$, we have

$$\left\| \tilde{L}_n(\Psi^k) \right\| \leq \left\| \tilde{L}_n(e_0) \right\|^{\frac{1}{p}} \left\| \tilde{L}_n(\Psi^{kq}) \right\|^{\frac{1}{q}},$$

which implies that

$$\left\| \tilde{L}_n(\Psi^k) \right\| \leq \left\| \tilde{L}_n(e_0) \right\|^{1 - \frac{k}{m+1}} \left\| \tilde{L}_n(\Psi^{m+1}) \right\|^{\frac{k}{m+1}} \quad (16.8)$$

for each $k = 1, 2, ..., m$. Using the fact that $|u + v|^\alpha \leq |u|^\alpha + |v|^\alpha$ for each $\alpha \in (0, 1]$, it follows from (16.8) that

$$\left\| \tilde{L}_n(\Psi^k) \right\| \leq \left\| \tilde{L}_n(e_0) - e_0 \right\|^{1 - \frac{k}{m+1}} \left\| \tilde{L}_n(\Psi^{m+1}) \right\|^{\frac{k}{m+1}} + \left\| \tilde{L}_n(\Psi^{m+1}) \right\|^{\frac{k}{m+1}}$$
(16.9)

for each $k = 1, 2, ..., m$. Combining (16.7) with (16.9) we derive that

$$D^* (L_n(f), f) \leq M \left\| \tilde{L}_n(e_0) - e_0 \right\| + M \sum_{k=1}^{m} \left\| \tilde{L}_n(\Psi^{m+1}) \right\|^{\frac{k}{m+1}}$$
$$+ M \sum_{k=1}^{m} \left\| \tilde{L}_n(e_0) - e_0 \right\|^{1 - \frac{k}{m+1}} \left\| \tilde{L}_n(\Psi^{m+1}) \right\|^{\frac{k}{m+1}}$$
$$+ M \left\| \tilde{L}_n(\Psi^{m+1}) \right\|^{\frac{m}{m+1}} \left\| \left(\tilde{L}_n(e_0) \right)^{\frac{1}{m+1}} + \frac{1}{m+1} \right\|$$
$$\times \omega_1^{(\mathcal{F})} \left(f^{(m)}, \left\| \tilde{L}_n(\Psi^{m+1}) \right\|^{\frac{1}{m+1}} \right),$$

where $M = \max\{M_0, M_1, ..., M_m, \frac{1}{m!}\}$. Hence, the above inequality yields that

$$\frac{1}{M}D^*(L_n(f), f) \leq \left\|\tilde{L}_n(e_0) - e_0\right\| + \sum_{k=1}^{m}\left\|\tilde{L}_n(\Psi^{m+1})\right\|^{\frac{k}{m+1}}$$

$$+ \sum_{k=1}^{m}\left\|\tilde{L}_n(e_0) - e_0\right\|^{1-\frac{k}{m+1}}\left\|\tilde{L}_n(\Psi^{m+1})\right\|^{\frac{k}{m+1}}$$

$$+ \left\|\tilde{L}_n(\Psi^{m+1})\right\|^{\frac{m}{m+1}}\left\|\tilde{L}_n(e_0) - e_0\right\|^{\frac{1}{m+1}}$$

$$\times \omega_1^{(\mathcal{F})}\left(f^{(m)}, \left\|\tilde{L}_n(\Psi^{m+1})\right\|^{\frac{1}{m+1}}\right)$$

$$+ \left\|\tilde{L}_n(\Psi^{m+1})\right\|^{\frac{m}{m+1}}\omega_1^{(\mathcal{F})}\left(f^{(m)}, \left\|\tilde{L}_n(\Psi^{m+1})\right\|^{\frac{1}{m+1}}\right)$$

and hence

$$D^*(L_n(f), f) \leq M\{p_n + q_n + r_n + s_n + t_n\}, \qquad (16.10)$$

where

$$p_n = \left\|\tilde{L}_n(e_0) - e_0\right\|,$$

$$q_n = \sum_{k=1}^{m}\left\|\tilde{L}_n(\Psi^{m+1})\right\|^{\frac{k}{m+1}},$$

$$r_n = \sum_{k=1}^{m}\left\|\tilde{L}_n(e_0) - e_0\right\|^{1-\frac{k}{m+1}}\left\|\tilde{L}_n(\Psi^{m+1})\right\|^{\frac{k}{m+1}},$$

$$s_n = \left\|\tilde{L}_n(\Psi^{m+1})\right\|^{\frac{m}{m+1}}\left\|\tilde{L}_n(e_0) - e_0\right\|^{\frac{1}{m+1}}\omega_1^{(\mathcal{F})}\left(f^{(m)}, \left\|\tilde{L}_n(\Psi^{m+1})\right\|^{\frac{1}{m+1}}\right),$$

$$t_n = \left\|\tilde{L}_n(\Psi^{m+1})\right\|^{\frac{m}{m+1}}\omega_1^{(\mathcal{F})}\left(f^{(m)}, \left\|\tilde{L}_n(\Psi^{m+1})\right\|^{\frac{1}{m+1}}\right).$$

Using the hypotheses (16.5), (16.6), and also considering Lemma 16.1, we get

$$st_A - \lim_n p_n = st_A - \lim_n q_n = st_A - \lim_n r_n = st_A - \lim_n s_n = st_A - \lim_n t_n = 0$$
$$(16.11)$$

On the other hand, by (16.10), we have, for every $\varepsilon > 0$ and each $j \in \mathbb{N}$, that

$$\sum_{n:D^*(L_n(f),f)\geq\varepsilon} a_{jn} \leq \sum_{n:p_n\geq\frac{\varepsilon}{5M}} a_{jn} + \sum_{n:q_n\geq\frac{\varepsilon}{5M}} a_{jn} + \sum_{n:r_n\geq\frac{\varepsilon}{5M}} a_{jn} + \sum_{n:s_n\geq\frac{\varepsilon}{5M}} a_{jn} + \sum_{n:t_n\geq\frac{\varepsilon}{5M}} a_{jn}.$$
$$(16.12)$$

Now taking limit as $j \to \infty$ and using (16.11) we observe that

$$\lim_j \sum_{n:D^*(L_n(f),f)\geq\varepsilon} a_{jn} = 0,$$

which finishes the proof.   ∎

16.2 Conclusions

If we replace the matrix A in Theorem 16.2 by the identity matrix, then we immediately obtain the next theorem which was first proved by Anastassiou in [5].

Corollary 16.3 (see [5]). *Let (L_n) be a sequence of fuzzy positive linear operators from $C_{\mathcal{F}}^{(m)}[a,b]$ into $C_{\mathcal{F}}[a,b]$. Assume that there exists a corresponding sequence $(\tilde{L}_n)$ of positive linear operators from $C^m[a,b]$ into $C[a,b]$ with the property (16.4). If the sequence $\left(\tilde{L}_n(e_0)\right)$ is uniformly convergent (in the ordinary sense) to the unit function e_0, and if the sequence $\left(\tilde{L}_n(\Psi^{m+1})\right)$ is uniformly convergent (in the ordinary sense) to the zero function on the interval $[a,b]$, then, for all $f \in C_{\mathcal{F}}^{(m)}[a,b]$, the sequence $(L_n(f))$ is uniformly convergent to f on $[a,b]$ (in the fuzzy sense).*

However, we can construct a sequence of fuzzy positive linear operators, which satisfies the statistical fuzzy approximation result (Theorem 16.2), but not Corollary 16.3. To see this let $A = [a_{jn}]$ be any non-negative regular summability matrix. Assume that K is any subset of $\mathbb{N}$ satisfying $\delta_A(K) = 0$. Then define a sequence (u_n) by:

$$u_n = \begin{cases} n^2, & \text{if } n \in K \\ \frac{1}{n^2}, & \text{if } n \in \mathbb{N}\backslash K. \end{cases} \tag{16.13}$$

Then define the fuzzy Bernstein-type polynomials as follows:

$$B_n^{\mathcal{F}}(f;x) = (1+u_n) \odot \bigoplus_{k=0}^n \binom{n}{k} x^k(1-x)^{n-k} \odot f\left(\frac{k}{n}\right),$$

where $f \in C_{\mathcal{F}}[0,1]$, $x \in [0,1]$ and $n \in \mathbb{N}$. We see that

$$\{B_n^{\mathcal{F}}(f;x)\}_{\pm}^{(r)} = \tilde{B}_n\left(f_{\pm}^{(r)};x\right) = (1+u_n)\sum_{k=0}^n \binom{n}{k} x^k(1-x)^{n-k} f_{\pm}^{(r)}\left(\frac{k}{n}\right),$$

where $f_{\pm}^{(r)} \in C[0,1]$, and we easily obtain that

$$\left\|\tilde{B}_n(e_0) - e_0\right\| = u_n \text{ with } e_0(y) := 1,$$

$$\left\|\tilde{B}_n\left(\Psi^2\right)\right\| = \frac{1+u_n}{4n} \text{ with } \Psi(y) = |y-x| \text{ for each } x \in [0,1].$$

Since $\delta_A(K) = 0$, we can write that

$$\sum_{n:|u_n|\geq\varepsilon} a_{jn} = \sum_{n\in K} a_{jn} \to 0 \ (\text{as } j \to \infty),$$

which implies

$$st_A - \lim_n u_n = 0.$$

Then we observe that

$$st_A - \lim_n \left\|\tilde{B}_n(e_0) - e_0\right\| = 0$$

and

$$st_A - \lim_n \left\|\tilde{B}_n\left(\Psi^2\right)\right\| = 0.$$

Hence all hypotheses of Theorem 16.2 hold for $m = 1$. Then, for all $f \in C^1_{\mathcal{F}}[0,1]$, we get

$$st_A - \lim_n \left\|B_n^{(\mathcal{F})}(f) - f\right\| = 0.$$

However, since the sequence (u_n) given by (16.13) is non-convergent (in the usual sense), for any $f \in C^1_{\mathcal{F}}[0,1]$, the sequence $\left(B_n^{(\mathcal{F})}(f)\right)$ is not fuzzy convergent to f.

Let $A = [a_{jn}]$ be a non-negative regular summability matrix and let (p_n) be a positive non-increasing sequence of real numbers. Then, we know from [62] that a sequence (x_n) is A-statistically convergent to the number L with the rate of $o(p_n)$ if for every $\varepsilon > 0$, $\lim_j \frac{1}{p_j} \sum_{n:|x_n-L|\geq\varepsilon} a_{jn} = 0$, which is denoted by $x_n - L = st_A - o(p_n)$ as $n \to \infty$. If, for every $\varepsilon > 0$, $\sup_j \frac{1}{p_j} \sum_{n:|x_n|\geq\varepsilon} a_{jn} < \infty$, then (x_n) is A-statistically bounded with the rate of $O(p_n)$, which is denoted by $x_n = st_A - O(p_n)$ as $n \to \infty$.

Using the above definitions the following auxiliary lemma was proved in [62].

Lemma 16.4 (see [62]). *Let (x_n) and (y_n) be two sequences. Assume that $A = [a_{jn}]$ is a non-negative regular summability matrix. Let (p_n) and (q_n) are positive non-increasing sequences. If for some real numbers L_1, L_2, the conditions $x_n - L_1 = st_A - o(p_n)$ and $y_n - L_2 = st_A - o(q_n)$ hold as $n \to \infty$, then we get*

$$(x_n - L_1) \pm (y_n - L_2) = st_A - o(r_n) \quad \text{as } n \to \infty$$

and

$$(x_n - L_1)(y_n - L_2) = st_A - o(r_n) \quad \text{as } n \to \infty$$

where $r_n := \max\{p_n, q_n\}$ for each $n \in \mathbb{N}$. Furthermore, similar results hold when little "o" is replaced by big "O".

On the other hand, the following result can be proved easily as in proof of Lemma 16.1.

Lemma 16.5. *Let* $A = [a_{jn}]$ *be a non-negative regular summability matrix, and let* (p_n) *be a positive non-increasing sequence of real numbers. If* (δ_n) *is a sequence of positive real numbers satisfying* $\delta_n = st_A - o(p_n)$ *as* $n \to \infty$, *then we have* $\omega_1^{(\mathcal{F})}(f, \delta_n) = st_A - o(p_n)$ *for all* $f \in C_{\mathcal{F}}[a, b]$.

Therefore, using Lemmas 16.4-16.5, and also considering inequality (16.10) one can obtain the next result at once, which shows the A-statistical rates of the approximation of fuzzy positive linear operators in Theorem 16.2.

Theorem 16.6. *Let* $A = [a_{jn}]$ *be a non-negative regular summability matrix and let* (L_n) *be a sequence of fuzzy positive linear operators from* $C_{\mathcal{F}}^{(m)}[a, b]$ *into* $C_{\mathcal{F}}[a, b]$. *Assume that there exists a corresponding sequence* $(\tilde{L}_n)$ *of positive linear operators from* $C^m[a, b]$ *into* $C[a, b]$ *with the property* (16.4). *Suppose that* (p_n) *and* (q_n) *are positive non-increasing sequences and also that the operators* $\tilde{L}_n$ *satisfy the following conditions:*

(i) $\left\| \tilde{L}_n(e_0) - e_0 \right\| = st_A - o(p_n)$ *as* $n \to \infty$, *where* $e_0(y) := 1$.

(ii) $\left\| \tilde{L}_n(\Psi^{m+1}) \right\| = st_A - o(q_n)$ *as* $n \to \infty$, *where* $\Psi(y) := |y - x|$ *for each* $x \in [a, b]$.

Then, for all $f \in C_{\mathcal{F}}^{(m)}[a, b]$, *we get*

$$D^* (L_n(f), f) = st_A - o(r_n) \text{ as } n \to \infty,$$

where $r_n := \max\{p_n, q_n\}$ *for each* $n \in \mathbb{N}$. *Furthermore, similar results hold when little "o" is replaced by big "O".*

17

Statistical Approximation by Bivariate Complex Picard Integral Operators

In this chapter, we investigate some statistical approximation properties of the bivariate complex Picard integral operators. Furthermore, we show that the statistical approach is more applicable than the well-known aspects. This chapter relies on [24].

17.1 Definition and Geometric Properties of the Operators

In this section, we mainly consider the idea as in the papers [37, 83]. Let

$$D^2 := D \times D = \left\{ (z, w) \in \mathbb{C}^2 : |z| < 1 \text{ and } |w| < 1 \right\}$$

and

$$\bar{D}^2 := \bar{D} \times \bar{D} = \left\{ (z, w) \in \mathbb{C}^2 : |z| \leq 1 \text{ and } |w| \leq 1 \right\}.$$

Assume that $f : \bar{D}^2 \to \mathbb{C}$ is a complex function in two complex variables. If the univariate complex functions $f(\cdot, w)$ and $f(z, \cdot)$ (for each fixed z and $w \in D$, respectively) are analytic on D, then we say that the function $f(\cdot, \cdot)$ is analytic on D^2 (see, e.g., [85, 94]). If a function f is analytic on D^2, then f has the following Taylor expansion

$$f(z, w) = \sum_{k,m=0}^{\infty} a_{k,m}(f) z^k w^m, \quad (z, w) \in D^2, \qquad (17.1)$$

G.A. Anastassiou and O. Duman: Towards Intelligent Modeling, ISRL 14, pp. 207–215.
springerlink.com   © Springer-Verlag Berlin Heidelberg 2011

having the $a_{k,m}(f)$ given by

$$a_{k,m}(f) := -\frac{1}{4\pi^2} \int_T \frac{f(p,q)}{p^{k+1}q^{m+1}} dpdq, \ k,m \in \mathbb{N}_0, \tag{17.2}$$

where $T := \{(p,q) \in \mathbb{C}^2 : |p| = r \text{ and } |q| = \rho\}$ with $0 < r, \rho < 1$.

Now consider the following space:

$$A(\bar{D}^2) := \{f : \bar{D}^2 \to \mathbb{C}; \ f \text{ is analytic on } D^2, \text{ continuous on } \bar{D}^2 \text{ with } f(0,0) = 0\}. \tag{17.3}$$

In this case, $A(\bar{D}^2)$ is a Banach space with the sup-norm given by

$$\|f\| = \sup\{|f(z,w)| : (z,w) \in \bar{D}^2\} \quad \text{for} \quad f \in A(\bar{D}^2).$$

We now define the bivariate complex Picard-type singular operators as follows:

$$P_n(f;z,w) := \frac{1}{2\pi\xi_n^2} \int_{-\infty}^{\infty} \int_{-\infty}^{\infty} f(ze^{is}, we^{it}) e^{-\sqrt{s^2+t^2}/\xi_n} dsdt, \tag{17.4}$$

where $(z,w) \in \bar{D}^2$, $n \in \mathbb{N}$, $f \in A(\bar{D}^2)$, and also (ξ_n) is a bounded sequence of positive real numbers.

It is easy to check that if f is a constant function on $\bar{D}^2$, say $f(z,w) \equiv C$, then we get, for every $n \in \mathbb{N}$ that $P_n(C;z,w) = C$. Hence, the operators P_n preserve the constant functions. In order to obtain some geometric properties of the operators P_n in (17.4) we need the following concepts.

Let $f \in C(\bar{D}^2)$, the space of all continuous functions on $\bar{D}^2$. Then, the first modulus of continuity of f on $\bar{D}^2$ denoted by $\omega_1(f,\delta)_{\bar{D}^2}$, $\delta > 0$, is defined to be

$$\omega_1(f;\delta)_{\bar{D}^2} := \sup\left\{|f(z,w) - f(p,q)| : \sqrt{|z-p|^2 + |w-q|^2} \le \delta, \ (z,w),(p,q) \in \bar{D}^2\right\}$$

and the second modulus of smoothness of f on $\partial(D^2)$ denoted by $\omega_2(f;\alpha)_{\partial(D^2)}$, $\alpha > 0$, is defined to by

$$\omega_2(f;\alpha)_{\partial(D^2)} := \sup\{f(e^{i(x+s)}, e^{i(y+t)}) - 2f(e^{ix}, e^{iy}) + f(e^{i(x-s)}, e^{i(y-t)}) : (x,y) \in \mathbb{R}^2 \text{ and } \sqrt{s^2+t^2} \le \alpha\}.$$

Then, by the maximum modulus principle for complex functions of several variables (see, e.g., [85, 94]), if $\sqrt{s^2+t^2} \le \alpha$, we see that

$$\left|f(ze^{is}, we^{it}) - 2f(z,w) + f(ze^{-is}, we^{-it})\right|$$
$$\le \sup_{(z,w)\in\bar{D}^2} \left|f(ze^{is}, we^{it}) - 2f(z,w) + f(ze^{-is}, we^{-it})\right|$$
$$= \sup_{(z,w)\in\partial(D^2)} \left|f(ze^{is}, we^{it}) - 2f(z,w) + f(ze^{-is}, we^{-it})\right|$$
$$= \sup_{(x,y)\in\mathbb{R}^2} \left|f(e^{i(x+s)}, e^{i(y+t)}) - 2f(e^{ix}, e^{iy}) + f(e^{i(x-s)}, e^{i(y-t)})\right|.$$

Thus, we easily obtain that

$$\left| f\left(ze^{is}, we^{it}\right) - 2f(z,w) + f\left(ze^{-is}, we^{-it}\right)\right| \le w_2\left(f; \sqrt{s^2+t^2}\right)_{\partial(D^2)}$$
(17.5)

Now let $f \in C\left(\bar{D}^2\right)$ and $\alpha > 0$. Using the function $\varphi_f : \mathbb{R}^2 \to \mathbb{C}$ given by $\varphi_f(x,y) = f\left(e^{ix}, e^{iy}\right)$, we observe that

$$w_2(f;\alpha)_{\partial(D^2)} \equiv w_2(\varphi_f;\alpha).$$
(17.6)

Therefore, the equivalence in (17.6) gives that

$$w_2(f;c\alpha)_{\partial(D^2)} \le (1+c)^2 w_2(f;\alpha)_{\partial(D^2)}.$$
(17.7)

We have the next result.

Theorem 17.1. *For each fixed $n \in \mathbb{N}$, we have $P_n\left(A\left(\bar{D}^2\right)\right) \subset A\left(\bar{D}^2\right)$.*

Proof. Let $n \in \mathbb{N}$ and $f \in A\left(\bar{D}^2\right)$ be fixed. Since $f(0,0) = 0$, we easily observe that

$$P_n(f;0,0) = \frac{1}{2\pi\xi_n^2} \int_{-\infty}^{\infty} \int_{-\infty}^{\infty} f(0,0)\, e^{-\sqrt{s^2+t^2}/\xi_n}\, ds\, dt = 0.$$

Now we prove that $P_n(f)$ is continuous on $\bar{D}^2$. To see this suppose that $(p,q), (z_m, w_m) \in \bar{D}^2$ and that $\lim_m(z_m, w_m) = (p,q)$. Hence, we obtain from the definition of w_1 that

$$|P_n(f; z_m, w_m) - P_n(f; p,q)|$$

$$\le \frac{1}{2\pi\xi_n^2} \int_{-\infty}^{\infty} \int_{-\infty}^{\infty} \left| f\left(z_m e^{is}, w_m e^{it}\right) - f\left(pe^{is}, qe^{it}\right)\right| e^{-\sqrt{s^2+t^2}/\xi_n}\, ds\, dt$$

$$\le \frac{w_1\left(f, \sqrt{|z_m - p|^2 + |w_m - q|^2}\right)_{\bar{D}^2}}{2\pi\xi_n^2} \int_{-\infty}^{\infty} \int_{-\infty}^{\infty} e^{-\sqrt{s^2+t^2}/\xi_n}\, ds\, dt$$

$$= w_1\left(f, \sqrt{|z_m - p|^2 + |w_m - q|^2}\right)_{\bar{D}^2}.$$

Since $\lim_m(z_m, w_m) = (p,q)$, we can write that

$$\lim_m \sqrt{|z_m - p|^2 + |w_m - q|^2} = 0,$$

which yields that

$$\lim_m w_1\left(f, \sqrt{|z_m - p|^2 + |w_m - q|^2}\right)_{\bar{D}^2} = 0.$$

due to the right continuity of $\omega_1(f,\cdot)_{\bar{D}^2}$ at zero. Hence, we get

$$\lim_m P_n(f; z_m, w_m) = P_n(f; p, q),$$

which gives the continuity of $P_n(f)$ at the point $(p, q) \in \bar{D}^2$.

Finally, since $f \in A(\bar{D}^2)$, the function f has the Taylor expansion in (17.1) with the coefficients $a_{k,m}(f)$ in (17.2). Then, for $(z, w) \in D^2$, we have

$$f(ze^{is}, we^{it}) = \sum_{k,m=0}^{\infty} a_{k,m}(f) z^k w^m e^{i(sk+tm)}. \tag{17.8}$$

Since $\left|a_{k,m}(f)e^{i(sk+tm)}\right| = |a_{k,m}(f)|$ for every $(s,t) \in \mathbb{R}^2$, the series in (17.8) is uniformly convergent with respect to $(s,t) \in \mathbb{R}^2$. Hence, we deduce that

$$
\begin{aligned}
P_n(f; z, w) &= \frac{1}{2\pi\xi_n^2} \int_{-\infty}^{\infty}\int_{-\infty}^{\infty} \left(\sum_{k,m=0}^{\infty} a_{k,m}(f) z^k w^m e^{i(sk+tm)}\right) e^{-\sqrt{s^2+t^2}/\xi_n}\, dsdt \\
&= \frac{1}{2\pi\xi_n^2} \sum_{k,m=0}^{\infty} a_{k,m}(f) z^k w^m \left(\int_{-\infty}^{\infty}\int_{-\infty}^{\infty} e^{i(sk+tm)} e^{-\sqrt{s^2+t^2}/\xi_n}\, dsdt\right) \\
&= \frac{1}{2\pi\xi_n^2} \sum_{k,m=0}^{\infty} a_{k,m}(f) z^k w^m \left(\int_{-\infty}^{\infty}\int_{-\infty}^{\infty} \cos(sk+tm) e^{-\sqrt{s^2+t^2}/\xi_n}\, dsdt\right) \\
&= \frac{2}{\pi\xi_n^2} \sum_{k,m=0}^{\infty} a_{k,m}(f) z^k w^m \left(\int_{0}^{\infty}\int_{0}^{\infty} \cos(sk+tm) e^{-\sqrt{s^2+t^2}/\xi_n}\, dsdt\right) \\
&= \sum_{k,m=0}^{\infty} a_{k,m}(f)\ell_n(k,m) z^k w^m,
\end{aligned}
$$

where, for $k, m \in \mathbb{N}_0$,

$$
\begin{aligned}
\ell_n(k,m) &:= \frac{2}{\pi\xi_n^2} \int_0^{\infty}\int_0^{\infty} \cos(sk+tm) e^{-\sqrt{s^2+t^2}/\xi_n}\, dsdt \\
&= \frac{2}{\pi\xi_n^2} \int_0^{\pi/2}\int_0^{\infty} \cos[\rho(k\cos\theta + m\sin\theta)] e^{-\rho/\xi_n} \rho\, d\rho d\theta \tag{17.9} \\
&= \frac{2}{\pi} \int_0^{\pi/2}\int_0^{\infty} \cos[u\xi_n(k\cos\theta + m\sin\theta)] e^{-u}\, du d\theta.
\end{aligned}
$$

We should note that

$$|\ell_n(k,m)| \leq 1 \text{ for every } n \in \mathbb{N} \text{ and } k, m \in \mathbb{N}_0.$$

Therefore, for each $n \in \mathbb{N}$ and $f \in A(\bar{D}^2)$, the function $P_n(f)$ has a Taylor series expansion whose Taylor coefficients are given by

$$a_{k,m}(P_n(f)) := a_{k,m}(f)\ell_n(k,m), \quad k, m \in \mathbb{N}_0. \tag{17.10}$$

Combining the above facts the proof is completed.   ∎

Now consider the following space:

$$B\left(\bar{D}^2\right) := \{f : \bar{D}^2 \to \mathbb{C};\ f \text{ is analytic on } D^2,\ f(0,0) = 1 \text{ and}$$
$$\operatorname{Re}\left[f(z,w)\right] > 0 \text{ for every } (z,w) \in D^2\}.$$

Then we get the following result.

Theorem 17.2. *For each fixed* $n \in \mathbb{N}$*, we have* $P_n\left(B\left(\bar{D}^2\right)\right) \subset B\left(\bar{D}^2\right).$

Proof. Let $n \in \mathbb{N}$ and $f \in B\left(\bar{D}^2\right)$ be fixed. As in the proof of Theorem 17.1, we observe that $P_n\left(f\right)$ is analytic on D^2. Since $f(0,0) = 1$, we easily see that

$$P_n(f;0,0) = \frac{1}{2\pi\xi_n^2} \int\limits_{-\infty}^{\infty} \int\limits_{-\infty}^{\infty} f(0,0)\, e^{-\sqrt{s^2+t^2}/\xi_n}\, ds dt = 1.$$

Finally, we can write that, for every $(z,w) \in D^2$,

$$\operatorname{Re}\left[P_n(f;z,w)\right] = \frac{1}{2\pi\xi_n^2} \int\limits_{-\infty}^{\infty} \int\limits_{-\infty}^{\infty} \operatorname{Re}\left[f\left(ze^{is}, we^{it}\right)\right] e^{-\sqrt{s^2+t^2}/\xi_n}\, ds dt > 0$$

since $\operatorname{Re}\left[f(z,w)\right] > 0$. Thus, the proof is done.   ∎

Using the definition of $\omega_1(f;\delta)_{\bar{D}^2}$ for $f \in C\left(\bar{D}^2\right)$ and $\delta > 0$, we obtain the next theorem.

Theorem 17.3. *For each fixed* $n \in \mathbb{N}$ *and* $f \in C\left(\bar{D}^2\right)$*, we get*

$$\omega_1(P_n(f);\delta)_{\bar{D}^2} \le \omega_1(f;\delta)_{\bar{D}^2}.$$

Proof. Let $\delta > 0$, $n \in \mathbb{N}$ and $f \in C\left(\bar{D}^2\right)$ be given. Assume that $(z,w),(p,q) \in \bar{D}^2$ and $\sqrt{|z-p|^2 + |w-q|^2} \le \delta$. Then, we get

$$|P_n\left(f;z,w\right) - P_n\left(f;p,q\right)|$$

$$\le \frac{1}{2\pi\xi_n^2} \int\limits_{-\infty}^{\infty} \int\limits_{-\infty}^{\infty} |f\left(ze^{is}, we^{it}\right) - f\left(pe^{is}, qe^{it}\right)|\, e^{-\sqrt{s^2+t^2}/\xi_n}\, ds dt$$

$$\le \omega_1\left(f; \sqrt{|z-p|^2 + |w-q|^2}\right)_{\bar{D}^2}$$

$$\le \omega_1\left(f;\delta\right)_{\bar{D}^2}.$$

Then, taking supremum over $\sqrt{|z-p|^2 + |w-q|^2} \le \delta$, we derive that

$$\omega_1(P_n(f);\delta)_{\bar{D}^2} \le \omega_1(f;\delta)_{\bar{D}^2},$$

which completes the proof.   ∎

17.2   Statistical Approximation of the Operators

We first give the next estimation for the operators P_n defined by (17.4).

Theorem 17.4. *For every* $f \in A\left(\bar{D}^2\right)$, *we get*

$$\|P_n(f) - f\| \leq M\omega_2\left(f, \xi_n\right)_{\partial(D^2)}$$

for some (finite) positive constant M.

Proof. Let $(z, w) \in \bar{D}^2$ and $f \in A\left(\bar{D}^2\right)$ be fixed. Then, we see that

$$
\begin{aligned}
P_n(f; z, w) - f(z, w) &= \frac{1}{2\pi\xi_n^2} \int\limits_{-\infty}^{\infty} \int\limits_{-\infty}^{\infty} \left\{ f\left(ze^{is}, we^{it}\right) - f(z, w) \right\} e^{-\sqrt{s^2+t^2}/\xi_n} \, ds\, dt \\
&= \frac{1}{2\pi\xi_n^2} \int\limits_{0}^{\infty} \int\limits_{0}^{\infty} \left\{ f\left(ze^{is}, we^{it}\right) - f(z, w) \right\} e^{-\sqrt{s^2+t^2}/\xi_n} \, ds\, dt \\
&\quad + \frac{1}{2\pi\xi_n^2} \int\limits_{-\infty}^{0} \int\limits_{-\infty}^{0} \left\{ f\left(ze^{is}, we^{it}\right) - f(z, w) \right\} e^{-\sqrt{s^2+t^2}/\xi_n} \, ds\, dt \\
&\quad + \frac{1}{2\pi\xi_n^2} \int\limits_{-\infty}^{0} \int\limits_{0}^{\infty} \left\{ f\left(ze^{is}, we^{it}\right) - f(z, w) \right\} e^{-\sqrt{s^2+t^2}/\xi_n} \, ds\, dt \\
&\quad + \frac{1}{2\pi\xi_n^2} \int\limits_{0}^{\infty} \int\limits_{-\infty}^{0} \left\{ f\left(ze^{is}, we^{it}\right) - f(z, w) \right\} e^{-\sqrt{s^2+t^2}/\xi_n} \, ds\, dt.
\end{aligned}
$$

After some simple calculations, we get

$$
\begin{aligned}
&P_n(f; z, w) - f(z, w) \\
&= \frac{1}{2\pi\xi_n^2} \int\limits_{0}^{\infty} \int\limits_{0}^{\infty} \{ f\left(ze^{is}, we^{it}\right) - 2f(z, w) + f\left(ze^{-is}, we^{-it}\right) \} e^{-\sqrt{s^2+t^2}/\xi_n} \, ds\, dt \\
&\quad + \frac{1}{2\pi\xi_n^2} \int\limits_{-\infty}^{0} \int\limits_{0}^{\infty} \{ f\left(ze^{is}, we^{it}\right) - 2f(z, w) + f\left(ze^{-is}, we^{-it}\right) \} e^{-\sqrt{s^2+t^2}/\xi_n} \, ds\, dt.
\end{aligned}
$$

It follows from the property (17.5) that, for all $(z,w) \in \bar{D}^2$,

$$|P_n(f;z,w) - f(z,w)| \leq \frac{1}{2\pi\xi_n^2} \int_0^\infty \int_0^\infty \omega_2\left(f, \sqrt{s^2+t^2}\right)_{\partial(D^2)} e^{-\sqrt{s^2+t^2}/\xi_n} ds dt$$

$$+ \frac{1}{2\pi\xi_n^2} \int_{-\infty}^0 \int_0^\infty \omega_2\left(f, \sqrt{s^2+t^2}\right)_{\partial(D^2)} e^{-\sqrt{s^2+t^2}/\xi_n} ds dt$$

$$= \frac{1}{\pi\xi_n^2} \int_0^\infty \int_0^\infty \omega_2\left(f, \sqrt{s^2+t^2}\right)_{\partial(D^2)} e^{-\sqrt{s^2+t^2}/\xi_n} ds dt$$

$$= \frac{1}{\pi\xi_n^2} \int_0^\infty \int_0^\infty \omega_2\left(f, \frac{\sqrt{s^2+t^2}}{\xi_n}\xi_n\right)_{\partial(D^2)} e^{-\sqrt{s^2+t^2}/\xi_n} ds dt.$$

If we also use (17.7), then we observe that

$$|P_n(f;z,w) - f(z,w)| \leq \frac{\omega_2(f,\xi_n)_{\partial(D^2)}}{\pi\xi_n^2} \int_0^\infty \int_0^\infty \left(1 + \frac{\sqrt{s^2+t^2}}{\xi_n}\right)^2 e^{-\sqrt{s^2+t^2}/\xi_n} ds dt$$

$$= \frac{\omega_2(f,\xi_n)_{\partial(D^2)}}{\pi\xi_n^2} \int_0^{\pi/2} \int_0^\infty \left(1 + \frac{\rho}{\xi_n}\right)^2 \rho e^{-\rho/\xi_n} d\rho d\theta$$

$$= \frac{\omega_2(f,\xi_n)_{\partial(D^2)}}{2} \int_0^\infty (1+u)^2 \rho e^{-u} d\rho$$

$$= M\omega_2(f,\xi_n)_{\partial(D^2)},$$

where

$$M = \frac{1}{2} \int_0^\infty (1+u)^2 u e^{-u} du < \infty.$$

Taking supremum over $(z,w) \in \bar{D}^2$ on the last inequality, the proof is finished. ∎

In order to obtain a statistical approximation by the operators P_n we need the next lemma.

Lemma 17.5. *Let* $A = [a_{jn}]$, $j,n = 1,2,...,$ *be a non-negative regular summability matrix. If a bounded sequence* (ξ_n) *in (17.4) satisfies the condition*

$$st_A - \lim \xi_n = 0, \tag{17.11}$$

then we get, for all $f \in C\left(\bar{D}^2\right)$, *that*

$$st_A - \lim_n \omega_2(f;\xi_n)_{\partial(D^2)} = 0.$$

Proof. Let $f \in C\left(\bar{D}^2\right)$. Then, the proof immediately follows from (17.11) and the right continuity of $\omega_2(f; \cdot)_{\partial(D^2)}$ at zero. ∎

Now we can give the following statistical approximation theorem.

Theorem 17.6. Let $A = [a_{jn}]$ be a non-negative regular summability matrix. Assume that the sequence (ξ_n) is the same as in Lemma 17.5. Then, for every $f \in A\left(\bar{D}^2\right)$, we get

$$st_A - \lim_n \|P_n(f) - f\| = 0.$$

Proof. Let $f \in A\left(\bar{D}^2\right)$. Then, for a given $\varepsilon > 0$, we can write from Theorem 17.4 that

$$U := \left\{n \in \mathbb{N} : \|P_n(f) - f\| \geq \varepsilon\right\} \subseteq \left\{n \in \mathbb{N} : \omega_2(f; \xi_n)_{\partial(D^2)} \geq \frac{\varepsilon}{M}\right\} =: V,$$

where M is the positive constant as in Theorem 17.4. Thus, for every $j \in \mathbb{N}$, we have

$$\sum_{n \in U} a_{jn} \leq \sum_{n \in V} a_{jn}.$$

Now letting $j \to \infty$ in the both sides of the last inequality and also considering Lemma 17.5 we see that

$$\lim_j \sum_{n \in U} a_{jn} = 0,$$

which implies

$$st_A - \lim_n \|P_n(f) - f\| = 0.$$

So, the proof is done. ∎

If we take $A = C_1$, the Cesáro matrix of order one, in Theorem 17.6, then we immediately get the next result.

Corollary 17.7. Let (ξ_n) be a bounded sequence of positive real numbers for which

$$st - \lim_n \xi_n = 0$$

holds. Then, for every $f \in A\left(\bar{D}^2\right)$, we get

$$st - \lim_n \|P_n(f) - f\| = 0.$$

Of course, if we choose $A = I$, the identity matrix, in Theorem 17.6, then we obtain the next uniform approximation result.

Corollary 17.8. Let (ξ_n) be a null sequence of positive real numbers. Then, for every $f \in A\left(\bar{D}^2\right)$, the sequence $(P_n(f))$ is uniformly convergent to f on $\bar{D}^2$.

Finally, define the sequence (ξ_n) as follows:

$$\xi_n := \begin{cases} n, & \text{if } n = k^2, \ k = 1, 2, \ldots \\ \frac{1}{n}, & \text{otherwise.} \end{cases} \tag{17.12}$$

Then, we see that $st - \lim_n \xi_n = 0$. In this case, by Corollary 17.7 (i.e., Theorem 17.6 for $A = C_1$) we derive that

$$st - \lim_n \|P_n(f) - f\| = 0$$

for every $f \in A\left(\bar{D}^2\right)$. However, since the sequence (ξ_n) given by (17.12) is non-convergent, the uniform approximation to a function f by the operators $P_n(f)$ is impossible.

We remark that the statistical results are still valid when $\lim \xi_n = 0$ because every convergent sequence is A-statistically convergent, and so statistically convergent. But, as in the above example, the statistical approximation theorems still work although (ξ_n) is non-convergent. Therefore, we can say that the approach presented in this chapter is more applicable than the classical case.

18

Statistical Approximation by Bivariate Complex Gauss-Weierstrass Integral Operators

In this chapter, we present the complex Gauss-Weierstrass integral operators defined on a space of analytic functions in two variables on the Cartesian product of two unit disks. Then, we investigate some geometric properties and statistical approximation process of these operators. This chapter relies on [30].

18.1 Definition and Geometric Properties of the Operators

We consider the following sets:

$$D^2 := D \times D = \left\{(z, w) \in \mathbb{C}^2 : |z| < 1 \text{ and } |w| < 1\right\},$$
$$\bar{D}^2 := \bar{D} \times \bar{D} = \left\{(z, w) \in \mathbb{C}^2 : |z| \leq 1 \text{ and } |w| \leq 1\right\}.$$

As usual, let $C\left(\bar{D}^2\right)$ denote the space of all continuous functions on $\bar{D}^2$. We also consider the space

$$A_1\left(\bar{D}^2\right) := \{f \in C\left(\bar{D}^2\right) : f \text{ is analytic on } D^2 \text{ with } f(0, 0) = 0\}.$$

Then, it is well-known that $C\left(\bar{D}^2\right)$ and $A_1\left(\bar{D}^2\right)$ are Banach spaces endowed with the usual sup-norm given by $\|f\| = \sup\left\{|f(z, w)| : (z, w) \in \bar{D}^2\right\}$.

Assume that (ξ_n) is a sequence of positive real numbers. Considering the sequence (λ_n) defined by

$$\lambda_n := \frac{1}{\pi\left(1 - e^{-\pi^2/\xi_n^2}\right)}, \tag{18.1}$$

G.A. Anastassiou and O. Duman: Towards Intelligent Modeling, ISRL 14, pp. 217–224.
springerlink.com   © Springer-Verlag Berlin Heidelberg 2011

and also using the set $\mathbb{D} := \left\{ (s,t) \in \mathbb{R}^2 : s^2 + t^2 \le \pi^2 \right\}$, we introduce the bivariate complex Gauss-Weierstrass singular integral operators as follows:

$$W_n(f; z, w) = \frac{\lambda_n}{\xi_n^2} \iint_{\mathbb{D}} f\left(ze^{is}, we^{it}\right) e^{-(s^2+t^2)/\xi_n^2} ds dt, \qquad (18.2)$$

where $(z, w) \in \bar{D}^2$, $n \in \mathbb{N}$, $f \in A_1\left(\bar{D}^2\right)$, and $(\lambda_n)_{n \in \mathbb{N}}$ is given by (18.1). Then, one can show that the operators W_n preserve the constant functions.

In order to obtain some geometric properties of the operators W_n in (18.2) we need the following concepts:

$$\omega_1(f; \delta)_{\bar{D}^2} := \sup\left\{ |f(z,w) - f(p,q)| : \sqrt{|z-p|^2 + |w-q|^2} \le \delta, \ (z,w), (p,q) \in \bar{D}^2 \right\},$$

which is the first modulus of continuity of $f \in C\left(\bar{D}^2\right)$; and

$$\omega_2(f; \alpha)_{\partial(D^2)} := \sup\{ f\left(e^{i(x+s)}, e^{i(y+t)}\right) - 2f\left(e^{ix}, e^{iy}\right) + f\left(e^{i(x-s)}, e^{i(y-t)}\right) : \\ (x,y) \in \mathbb{R}^2 \text{ and } \sqrt{s^2 + t^2} \le \alpha\},$$

which is the second modulus of smoothness of $f \in \partial\left(D^2\right)$. Then, it is not hard to see that, for $\sqrt{s^2 + t^2} \le \alpha$, we get (see [24])

$$\left| f\left(ze^{is}, we^{it}\right) - 2f(z,w) + f\left(ze^{-is}, we^{-it}\right) \right| \le \omega_2\left(f; \sqrt{s^2 + t^2}\right)_{\partial(D^2)}. \qquad (18.3)$$

We can also write, for any $c, \alpha > 0$, that

$$\omega_2(f; c\alpha)_{\partial(D^2)} \le (1+c)^2 \omega_2(f; \alpha)_{\partial(D^2)}. \qquad (18.4)$$

Finally, we define the following set:

$$B\left(\bar{D}^2\right) := \{f : \bar{D}^2 \to \mathbb{C}; f \text{ is analytic on } D^2, f(0,0) = 1 \text{ and} \\ \text{Re}\left[f(z,w)\right] > 0 \text{ for every } (z,w) \in D^2\}.$$

Now, using the idea introduced in [37, 83], we get the next result.

Theorem 18.1. *For each fixed $n \in \mathbb{N}$, we get*

(i) $W_n\left(A_1\left(\bar{D}^2\right)\right) \subset A_1\left(\bar{D}^2\right)$,

(ii) $W_n\left(B\left(\bar{D}^2\right)\right) \subset B\left(\bar{D}^2\right)$,

(iii) $\omega_1(W_n(f); \delta)_{\bar{D}^2} \le \omega_1(f; \delta)_{\bar{D}^2}$ for any $\delta > 0$ and for every $f \in C\left(\bar{D}^2\right)$.

Proof. (i) Let $f \in A_1\left(\bar{D}^2\right)$. Then, we have $f(0,0) = 0$, and so $W_n(f; 0, 0) = 0$. Now we prove that $W_n(f)$ is continuous on $\bar{D}^2$. Indeed, if $(p, q), (z_m, w_m) \in \bar{D}^2$ and $\lim_m(z_m, w_m) = (p, q)$, then we obtain that

$$|W_n(f; z_m, w_m) - W_n(f; p, q)|$$

$$\leq \frac{\lambda_n}{\xi_n^2} \iint_{\mathbb{D}} \left| f\left(z_m e^{is}, w_m e^{it}\right) - f\left(p e^{is}, q e^{it}\right)\right| e^{-(s^2+t^2)/\xi_n^2} ds dt$$

$$\leq \frac{\lambda_n \omega_1\left(f, \sqrt{|z_m - p|^2 + |w_m - q|^2}\right)_{\bar{D}^2}}{\xi_n^2} \iint_{\mathbb{D}} e^{-(s^2+t^2)/\xi_n^2} ds dt$$

$$= \omega_1\left(f, \sqrt{|z_m - p|^2 + |w_m - q|^2}\right)_{\bar{D}^2}.$$

Since $\lim_m(z_m, w_m) = (p, q)$, we can write that $\lim_m \sqrt{|z_m - p|^2 + |w_m - q|^2} = 0$, which gives that

$$\lim_m \omega_1\left(f, \sqrt{|z_m - p|^2 + |w_m - q|^2}\right)_{\bar{D}^2} = 0.$$

due to the right continuity of $\omega_1(f, \cdot)_{\bar{D}^2}$ at zero. Hence, we derive $\lim_m W_n(f; z_m, w_m) = W_n(f; p, q)$, which implies the continuity of $W_n(f)$ at the point $(p, q) \in \bar{D}^2$. Since $f \in A_1\left(\bar{D}^2\right)$, the function f has the following Taylor expansion

$$f(z, w) = \sum_{k,m=0}^{\infty} a_{k,m}(f) z^k w^m, \quad (z, w) \in D^2$$

with the coefficients $a_{k,m}(f)$ given by

$$a_{k,m}(f) = -\frac{1}{4\pi^2} \int_T \frac{f(p, q)}{p^{k+1} q^{m+1}} dp dq, \quad k, m \in \mathbb{N}_0,$$

where $T := \left\{(p, q) \in \mathbb{C}^2 : |p| = r \text{ and } |q| = \rho\right\}$ with $0 < r, \rho < 1$ (see, for instance, [85, 94]), we obtain, for every $(z, w) \in D^2$, that

$$f(z e^{is}, w e^{it}) = \sum_{k,m=0}^{\infty} a_{k,m}(f) z^k w^m e^{i(sk+tm)}. \tag{18.5}$$

Since $\left|a_{k,m}(f) e^{i(sk+tm)}\right| = |a_{k,m}(f)|$ for every $(s, t) \in \mathbb{R}^2$, the series in (18.5) is uniformly convergent with respect to $(s, t) \in \mathbb{R}^2$. Hence, we deduce that

$$W_n(f;z,w) = \frac{\lambda_n}{\xi_n^2} \iint\limits_{\mathbb{D}} \left(\sum_{k,m=0}^{\infty} a_{k,m}(f) z^k w^m e^{i(sk+tm)} \right) e^{-(s^2+t^2)/\xi_n^2} ds dt$$

$$= \frac{\lambda_n}{\xi_n^2} \sum_{k,m=0}^{\infty} a_{k,m}(f) z^k w^m \left(\iint\limits_{\mathbb{D}} e^{i(sk+tm)} e^{-(s^2+t^2)/\xi_n^2} ds dt \right)$$

$$= \frac{\lambda_n}{\xi_n^2} \sum_{k,m=0}^{\infty} a_{k,m}(f) z^k w^m \left(\iint\limits_{\mathbb{D}} \cos(sk+tm) e^{-(s^2+t^2)/\xi_n^2} ds dt \right)$$

$$=: \sum_{k,m=0}^{\infty} a_{k,m}(f) \ell_n(k,m) z^k w^m,$$

where, for $k, m \in \mathbb{N}_0$,

$$
\begin{aligned}
\ell_n(k,m) &:= \frac{\lambda_n}{\xi_n^2} \iint\limits_{\mathbb{D}} \cos(sk+tm) e^{-(s^2+t^2)/\xi_n^2} ds dt \\
&= \frac{\lambda_n}{\xi_n^2} \int_0^{2\pi} \int_0^{\pi} \cos[\rho(k\cos\theta + m\sin\theta)] e^{-\rho^2/\xi_n^2} \rho d\rho d\theta.
\end{aligned}
\tag{18.6}
$$

We should note that

$$|\ell_n(k,m)| \leq 1 \text{ for every } n \in \mathbb{N} \text{ and } k, m \in \mathbb{N}_0.$$

Therefore, for each $n \in \mathbb{N}$ and $f \in A_1(\bar{D}^2)$, the function $W_n(f)$ has a Taylor series expansion whose Taylor coefficients are defined by

$$a_{k,m}(W_n(f)) := a_{k,m}(f) \ell_n(k,m), \quad k, m \in \mathbb{N}_0,$$

where $\ell_n(k,m)$ is given by (18.6). Combining the above facts we derive that $W_n(f) \in A_1(\bar{D}^2)$. Since $f \in A_1(\bar{D}^2)$ was arbitrary, we immediately see that $W_n(A_1(\bar{D}^2)) \subset A_1(\bar{D}^2)$.

(ii) Now let $f \in B(\bar{D}^2)$ be fixed. As in the proof of (i), we observe that $W_n(f)$ is analytic on D^2. Since $f(0,0) = 1$, we see that $W_n(f;0,0) = 1$. Also, since $\text{Re}[f(z,w)] > 0$ for every $(z,w) \in D^2$, we have

$$\text{Re}[W_n(f;z,w)] = \frac{\lambda_n}{\xi_n^2} \iint\limits_{\mathbb{D}} \text{Re}\left[f\left(ze^{is}, we^{it}\right)\right] e^{-(s^2+t^2)/\xi_n^2} ds dt > 0.$$

(iii) Let $\delta > 0$ and $f \in C(\bar{D}^2)$. Assume that $(z,w), (p,q) \in \bar{D}^2$ and

$$\sqrt{|z-p|^2 + |w-q|^2} \leq \delta.$$

Then, we get

$$|W_n\left(f;z,w\right) - W_n\left(f;p,q\right)|$$

$$\leq \frac{\lambda_n}{\xi_n^2} \iint\limits_{\mathbb{D}} \left|f\left(ze^{is},we^{it}\right) - f\left(pe^{is},qe^{it}\right)\right| e^{-(s^2+t^2)/\xi_n^2} ds dt$$

$$\leq \omega_1\left(f;\sqrt{|z-p|^2 + |w-q|^2}\right)_{\bar{D}^2}$$

$$\leq \omega_1\left(f;\delta\right)_{\bar{D}^2},$$

which implies

$$\omega_1(W_n(f);\delta)_{\bar{D}^2} \leq \omega_1(f;\delta)_{\bar{D}^2}.$$

Thus, the proof is finished.   ∎

18.2 Statistical Approximation of the Operators

In order to obtain some statistical approximation theorems we mainly use the concept of A-statistical convergence, where $A = [a_{jn}]$, $j,n = 1,2,...$, is any non-negative regular summability matrix. We first need the following estimation.

Lemma 18.2. *For every* $f \in A_1\left(\bar{D}^2\right)$, *we get*

$$\|W_n(f) - f\| \leq \frac{M}{1 - e^{-\pi^2/\xi_n^2}} \omega_2\left(f,\xi_n\right)_{\partial(D^2)}$$

for some (finite) positive constant M *independent from* n.

Proof. Let $(z,w) \in \bar{D}^2$ and $f \in A_1\left(\bar{D}^2\right)$ be fixed. Define the following subsets of the set $\mathbb{D}$:

$$\mathbb{D}_1 := \{(s,t) \in \mathbb{D} : s \geq 0, t \geq 0\},$$
$$\mathbb{D}_2 := \{(s,t) \in \mathbb{D} : s \leq 0, t \leq 0\},$$
$$\mathbb{D}_3 := \{(s,t) \in \mathbb{D} : s \leq 0, t \geq 0\},$$
$$\mathbb{D}_4 := \{(s,t) \in \mathbb{D} : s \geq 0, t \leq 0\},$$

Then, we see that

$$W_n(f; z, w) - f(z, w) = \frac{\lambda_n}{\xi_n^2} \iint_{\mathbb{D}} \left\{ f\left(ze^{is}, we^{it}\right) - f(z, w) \right\} e^{-(s^2+t^2)/\xi_n^2} ds dt$$

$$= \frac{\lambda_n}{\xi_n^2} \iint_{\mathbb{D}_1} \left\{ f\left(ze^{is}, we^{it}\right) - f(z, w) \right\} e^{-(s^2+t^2)/\xi_n^2} ds dt$$

$$+ \frac{\lambda_n}{\xi_n^2} \iint_{\mathbb{D}_2} \left\{ f\left(ze^{is}, we^{it}\right) - f(z, w) \right\} e^{-(s^2+t^2)/\xi_n^2} ds dt$$

$$+ \frac{\lambda_n}{\xi_n^2} \iint_{\mathbb{D}_3} \left\{ f\left(ze^{is}, we^{it}\right) - f(z, w) \right\} e^{-(s^2+t^2)/\xi_n^2} ds dt$$

$$+ \frac{\lambda_n}{\xi_n^2} \iint_{\mathbb{D}_4} \left\{ f\left(ze^{is}, we^{it}\right) - f(z, w) \right\} e^{-(s^2+t^2)/\xi_n^2} ds dt.$$

Thus, we get

$$W_n(f; z, w) - f(z, w)$$
$$= \frac{\lambda_n}{\xi_n^2} \iint_{\mathbb{D}_1} \{ f\left(ze^{is}, we^{it}\right) - 2f(z, w) + f\left(ze^{-is}, we^{-it}\right) \} e^{-(s^2+t^2)/\xi_n^2} ds dt$$

$$+ \frac{\lambda_n}{\xi_n^2} \iint_{\mathbb{D}_3} \{ f\left(ze^{is}, we^{it}\right) - 2f(z, w) + f\left(ze^{-is}, we^{-it}\right) \} e^{-(s^2+t^2)/\xi_n^2} ds dt.$$

The property (18.3) implies that

$$|W_n(f; z, w) - f(z, w)| \leq \frac{\lambda_n}{\xi_n^2} \iint_{\mathbb{D}_1} \omega_2 \left(f, \sqrt{s^2 + t^2} \right)_{\partial(D^2)} e^{-(s^2+t^2)/\xi_n^2} ds dt$$

$$+ \frac{\lambda_n}{\xi_n^2} \iint_{\mathbb{D}_3} \omega_2 \left(f, \sqrt{s^2 + t^2} \right)_{\partial(D^2)} e^{-(s^2+t^2)/\xi_n^2} ds dt$$

$$= \frac{2\lambda_n}{\xi_n^2} \iint_{\mathbb{D}_1} \omega_2 \left(f, \sqrt{s^2 + t^2} \right)_{\partial(D^2)} e^{-(s^2+t^2)/\xi_n^2} ds dt$$

$$= \frac{2\lambda_n}{\xi_n^2} \iint_{\mathbb{D}_1} \omega_2 \left(f, \frac{\sqrt{s^2 + t^2}}{\xi_n} \xi_n \right)_{\partial(D^2)} e^{-(s^2+t^2)/\xi_n^2} ds dt.$$

Also using (18.4), then we observe that

$$|W_n(f;z,w) - f(z,w)| \leq \frac{2\lambda_n \omega_2 (f,\xi_n)_{\partial(D^2)}}{\xi_n^2} \iint\limits_{\mathbb{D}_1} \left(1 + \frac{\sqrt{s^2+t^2}}{\xi_n}\right)^2 e^{-(s^2+t^2)/\xi_n^2} ds dt$$

$$= \frac{2\lambda_n \omega_2 (f,\xi_n)_{\partial(D^2)}}{\xi_n^2} \int_0^{\pi/2} \int_0^{\pi} \left(1 + \frac{\rho}{\xi_n}\right)^2 \rho e^{-\rho^2/\xi_n^2} d\rho d\theta$$

$$= \pi\lambda_n \omega_2 (f,\xi_n)_{\partial(D^2)} \int_0^{\pi} (1+u)^2 u e^{-u^2} du$$

$$= \frac{M}{1 - e^{-\pi^2/\xi_n^2}} \omega_2 (f,\xi_n)_{\partial(D^2)},$$

where

$$M = \int_0^{\pi} (1+u)^2 u e^{-u^2} du < \infty.$$

Taking supremum over $(z,w) \in \bar{D}^2$ on the last inequality, the proof is done. ∎

Then we obtain the following statistical approximation theorem.

Theorem 18.3. *Let $A = [a_{jn}]$ be a non-negative regular summability matrix. If (ξ_n) is a sequence of positive real numbers satisfying*

$$st_A - \lim_n \xi_n = 0, \tag{18.7}$$

then, for every $f \in A_1 (\bar{D}^2)$, we get

$$st_A - \lim_n \|W_n(f) - f\| = 0.$$

Proof. Let $f \in A_1 (\bar{D}^2)$. By (18.7), we easily observe that

$$st_A - \lim_n \frac{1}{1 - e^{-\pi^2/\xi_n^2}} = 0.$$

Then, from Lemma 17.5 obtain in Chapter 17, we can write that

$$st_A - \lim_n \frac{\omega_2(f;\xi_n)_{\partial(D^2)}}{1 - e^{-\pi^2/\xi_n^2}} = 0. \tag{18.8}$$

Hence, for a given $\varepsilon > 0$, it follows from Lemma 18.2 that

$$U := \{n \in \mathbb{N} : \|W_n(f) - f\| \geq \varepsilon\} \subseteq \left\{n \in \mathbb{N} : \frac{\omega_2(f;\xi_n)_{\partial(D^2)}}{1 - e^{-\pi^2/\xi_n^2}} \geq \frac{\varepsilon}{M}\right\} =: V,$$

where M is the positive constant as in the proof of Lemma 18.2. The last inclusion implies, for every $j \in \mathbb{N}$, that

$$\sum_{n \in U} a_{jn} \le \sum_{n \in V} a_{jn}.$$

Now taking limit as $j \to \infty$ and then using (18.8) we derive that

$$\lim_j \sum_{n \in U} a_{jn} = 0,$$

which gives

$$st_A - \lim_n \|W_n(f) - f\| = 0.$$

The proof is finished.   ∎

If we take $A = C_1$ in Theorem 18.3, then we easily obtain the next result.

Corollary 18.4. *Let (ξ_n) be a sequence of positive real numbers satisfying $st - \lim_n \xi_n = 0$, then, for every $f \in A_1(\bar{D}^2)$, we get $st - \lim_n \|W_n(f) - f\| = 0$.*

Of course, if we choose $A = I$, the identity matrix, in Theorem 18.3, then we have the next uniform approximation result.

Corollary 18.5. *Let (ξ_n) be a null sequence of positive real numbers. Then, for every $f \in A_1(\bar{D}^2)$, the sequence $(W_n(f))$ is uniformly convergent to f on $\bar{D}^2$.*

Finally, as in [24], if we take $A = C_1$, the Cesáro matrix of order one, and define the sequence (ξ_n) by

$$\xi_n := \begin{cases} \frac{n}{n+1}, & \text{if } n = k^2, \ k = 1, 2, \dots \\ 0, & \text{otherwise}, \end{cases} \tag{18.9}$$

then, the statistical approximation result in Corollary 18.4 (or, Theorem 18.3) works for the operators W_n constructed with the sequence (ξ_n) in (18.9), however the uniform approximation to a function $f \in A_1(\bar{D}^2)$ are impossible since (ξ_n) is a non-convergent sequence in the usual sense.

References

1. Altomare, F., Campiti, M.: Korovkin Type Approximation Theory and Its Application. Walter de Gruyter Publ., Berlin (1994)
2. Altomare, F., Rasa, I.: Approximation by positive operators in spaces $C^p([a,b])$. Anal. Numér. Théor. Approx. 18, 1–11 (1989)
3. Anastassiou, G.A.: Korovkin inequalities for stochastic processes. J. Math. Anal. Appl. 157, 366–384 (1991)
4. Anastassiou, G.A.: Fuzzy approximation by fuzzy convolution type operators. Comput. Math. Appl. 48, 1369–1386 (2004)
5. Anastassiou, G.A.: High-order fuzzy approximation by fuzzy wavelet type and neural network operators. Comput. Math. Appl. 48, 1387–1401 (2004)
6. Anastassiou, G.A.: On basic fuzzy Korovkin theory. Studia Univ. Babeş-Bolyai Math. 50, 3–10 (2005)
7. Anastassiou, G.A.: Global smoothness and uniform convergence of smooth Picard singular operators. Comput. Math. Appl. 50, 1755–1766 (2005)
8. Anastassiou, G.A.: L_p convergence with rates of smooth Picard singular operators. In: Differential & Difference Equations and Applications, pp. 31–45. Hindawi Publ. Corp., New York (2006)
9. Anastassiou, G.A.: Basic convergence with rates of smooth Picard singular integral operators. J. Comput. Anal. Appl. 8, 313–334 (2006)
10. Anastassiou, G.A.: Fuzzy random Korovkin theory and inequalities. Math. Inequal. Appl. 10, 63–94 (2007)
11. Anastassiou, G.A.: Stochastic Korovkin theory given quantitatively. Facta Univ. Ser. Math. Inform. 22, 43–60 (2007)
12. Anastassiou, G.A.: Multivariate stochastic Korovkin theory given quantitatively. Math. Comput. Modelling 48, 558–580 (2008)
13. Anastassiou, G.A.: Fractional Korovkin theory. Chaos, Solitons & Fractals 42, 2080–2094 (2009)

14. Anastassiou, G.A.: Fractional Differentiation Inequalities. Springer, New York (2009)
15. Anastassiou, G.A.: Fractional trigonometric Korovkin theory. Commun. Appl. Anal. 14, 39–58 (2010)
16. Anastassiou, G.A., Duman, O.: A Baskakov type generalization of statistical Korovkin theory. J. Math. Anal. Appl. 340, 476–486 (2008)
17. Anastassiou, G.A., Duman, O.: Statistical fuzzy approximation by fuzzy positive linear operators. Comput. Math. Appl. 55, 573–580 (2008)
18. Anastassiou, G.A., Duman, O.: On relaxing the positivity condition of linear operators in statistical Korovkin-type approximations. J. Comput. Anal. Appl. 11, 7–19 (2009)
19. Anastassiou, G.A., Duman, O.: Statistical weighted approximation to derivatives of functions by linear operator. J. Comput. Anal. Appl. 11, 20–30 (2009)
20. Anastassiou, G.A., Duman, O.: Statistical approximation to periodic functions by a general class of linear operators. J. Concr. Appl. Math. 7, 200–207 (2009)
21. Anastassiou, G.A., Duman, O.: Fractional Korovkin theory based on statistical convergence. Serdica Math. J. 35, 381–396 (2009)
22. Anastassiou, G.A., Duman, O.: High order statistical fuzzy Korovkin theory. Stochastic Anal. Appl. 27, 543–554 (2009)
23. Anastassiou, G.A., Duman, O.: Statistical L_p-approximation by double Gauss-Weierstrass singular integral operators. Comput. Math. Appl. 59, 1985–1999 (2010)
24. Anastassiou, G.A., Duman, O.: Statistical convergence of double complex Picard integral operators. Appl. Math. Lett. 23, 852–858 (2010)
25. Anastassiou, G.A., Duman, O.: Statistical approximation by double Picard singular integral operators. Studia Univ. Babeş-Bolyai Math. 55, 3–20 (2010)
26. Anastassiou, G.A., Duman, O.: Statistical Korovkin theory for multivariate stochastic processes. Stochastic Anal. Appl. 28, 648–661 (2010)
27. Anastassiou, G.A., Duman, O.: Fractional trigonometric Korovkin theory in statistical sense. Serdica Math. J. 36, 121–136 (2010)
28. Anastassiou, G.A., Duman, O.: Uniform approximation in statistical sense by double Gauss-Weierstrass singular integral operators. Nonlinear Funct. Anal. Appl. (accepted for publication)
29. Anastassiou, G.A., Duman, O.: Statistical L_p-convergence of double smooth Picard singular integral operators. J. Comput. Anal. Appl. (accepted for publication)
30. Anastassiou, G.A., Duman, O.: Statistical approximation by double complex Gauss-Weierstrass integral operators. Appl. Math. Lett. (accepted for publication)
31. Anastassiou, G.A., Duman, O., Erkuş-Duman, E.: Statistical approximation for stochastic processes. Stochastic Anal. Appl. 27, 460–474 (2009)
32. Anastassiou, G.A., Gal, S.G.: General theory of global smoothness preservation by singular integrals, univariate case. J. Comput. Anal. Appl. 1, 289–317 (1999)

33. Anastassiou, G.A., Gal, S.G.: Approximation Theory: Moduli of Continuity and Global Smoothness Preservation. Birkhäuser Boston, Inc., Boston (2000)
34. Anastassiou, G.A., Gal, S.G.: Convergence of generalized singular integrals to the unit, multivariate case. Applied Mathematics Reviews 1, 1–8 (2000)
35. Anastassiou, G.A., Gal, S.G.: Convergence of generalized singular integrals to the unit, univariate case. Math. Inequal. Appl. 4, 511–518 (2000)
36. Anastassiou, G.A., Gal, S.G.: On fuzzy trigonometric Korovkin theory. Nonlinear Funct. Anal. Appl. 11, 385–395 (2006)
37. Anastassiou, G.A., Gal, S.G.: Geometric and approximation properties of some singular integrals in the unit disk. J. Inequal. Appl., Art. ID 17231, 19 (2006)
38. Aral, A.: Pointwise approximation by the generalization of Picard and Gauss-Weierstrass singular integrals. J. Concr. Appl. Math. 6, 327–339 (2008)
39. Aral, A., Gal, S.G.: q-generalizations of the Picard and Gauss-Weierstrass singular integrals. Taiwanese J. Math. 12, 2501–2515 (2008)
40. Baskakov, V.A.: Generalization of certain theorems of P.P. Korovkin on positive operators for periodic functions (Russian). In: A collection of articles on the constructive theory of functions and the extremal problems of functional analysis (Russian), pp. 40–44. Kalinin Gos. Univ., Kalinin (1972)
41. Baskakov, V.A.: Generalization of certain theorems of P.P. Korovkin on positive operators (Russian). Math. Zametki 13, 785–794 (1973)
42. Boos, J.: Classical and Modern Methods in Summability. Oxford Univ. Press, UK (2000)
43. Brosowksi, B.: A Korovkin-type theorem for differentiable functions. In: Approximation Theory, III (Proc. Conf., Univ. Texas, Austin, Tex.), pp. 255–260. Academic Press, New York (1980)
44. Burgin, M., Duman, O.: Statistical fuzzy convergence. Internat. J. Uncertain. Fuzziness Knowledge-Based Systems 16, 879–902 (2008)
45. Connor, J.S.: The statistical and p-Cesáro convergence of sequences. Analysis 8, 47–63 (1988)
46. Connor, J.S.: Two valued measures and summability. Analysis 10, 373–385 (1990)
47. Connor, J.S.: R-type summability methods, Cauchy criteria, P-sets and statistical convergence. Proc. Amer. Math. Soc. 115, 319–327 (1992)
48. Connor, J.S., Ganichev, M., Kadets, V.: A characterization of Banach spaces with separable duals via weak statistical convergence. J. Math. Anal. Appl. 244, 251–261 (2000)
49. Connor, J.S., Kline, J.: On statistical limit points and the consistency of statistical convergence. J. Math. Anal. Appl. 197, 393–399 (1996)
50. Connor, J.S., Swardson, M.A.: Strong integral summability and the Stone-Čhech compactification of the half-line. Pacific J. Math. 157, 201–224 (1993)
51. Delgado, F.J.M., Gonzáles, V.R., Morales, D.C.: Qualitative Korovkin-type results on conservative approximation. J. Approx. Theory 94, 144–159 (1998)
52. Demirci, K.: A-statistical core of a sequence. Demonstratio Math. 33, 343–353 (2000)

53. Devore, R.A.: The Approximation of Continuous Functions by Positive Linear Operators. Lecture Notes in Mathematics, vol. 293. Springer, Berlin (1972)
54. Doğru, O., Duman, O.: Statistical approximation of Meyer-König and Zeller operators based on q-integers. Publ. Math. Debrecen. 68, 199–214 (2006)
55. Doğru, O., Duman, O., Orhan, C.: Statistical approximation by generalized Meyer-König and Zeller type operators. Studia Sci. Math. Hungar. 40, 359–371 (2003)
56. Duman, O.: Statistical approximation for periodic functions. Demonstratio Math. 36, 873–878 (2003)
57. Duman, O.: Statistical approximation theorems by k -positive linear operators. Arch. Math (Basel) 86, 569–576 (2006)
58. Duman, O.: Regular matrix transformations and rates of convergence of positive linear operators. Calcolo 44, 159–164 (2007)
59. Duman, O., Anastassiou, G.A.: On statistical fuzzy trigonometric Korovkin theory. J. Comput. Anal. Appl. 10, 333–344 (2008)
60. Duman, O., Erkuş, E.: Approximation of continuous periodic functions via statistical convergence. Comput. Math. Appl. 52, 967–974 (2006)
61. Duman, O., Erkuş, E., Gupta, V.: Statistical rates on the multivariate approximation theory. Math. Comput. Modelling 44, 763–770 (2006)
62. Duman, O., Khan, M.K., Orhan, C.: A-Statistical convergence of approximating operators. Math. Inequal. Appl. 6, 689–699 (2003)
63. Duman, O., Orhan, C.: Statistical approximation by positive linear operators. Studia. Math. 161, 187–197 (2004)
64. Duman, O., Orhan, C.: An abstract version of the Korovkin approximation theorem. Publ. Math. Debrecen. 69, 33–46 (2006)
65. Duman, O., Özarslan, M.A., Doğru, O.: On integral type generalizations of positive linear operators. Studia Math. 174, 1–12 (2006)
66. Efendiev, R.O.: Conditions for convergence of linear operators to derivatives (Russian). Akad. Nauk Azerb. SSR Dokl. 40, 3–6 (1984)
67. Erkuş, E., Duman, O.: A-statistical extension of the Korovkin type approximation theorem. Proc. Indian Acad. Sci (Math. Sci.) 115, 499–508 (2005)
68. Erkuş, E., Duman, O.: A Korovkin type approximation theorem in statistical sense. Studia Sci. Math. Hungar. 43, 285–294 (2006)
69. Erkuş, E., Duman, O., Srivastava, H.M.: Statistical approximation of certain positive linear operators constructed by means of the Chan-Chyan-Srivastava polynomials. Appl. Math. Comput. 182, 213–222 (2006)
70. Fast, H.: Sur la convergence statistique. Colloq. Math. 2, 241–244 (1951)
71. Freedman, A.R., Sember, J.J.: Densities and summability. Pacific J. Math. 95, 293–305 (1981)
72. Fridy, J.A.: Minimal rates of summability. Canad. J. Math. 30, 808–816 (1978)
73. Fridy, J.A.: On statistical convergence. Analysis 5, 301–313 (1985)
74. Fridy, J.A., Khan, M.K.: Tauberian theorems via statistical convergence. J. Math. Anal. Appl. 228, 73–95 (1998)
75. Fridy, J.A., Miller, H.I., Orhan, C.: Statistical rates of convergence. Acta Sci. Math. (Szeged) 69, 147–157 (2003)
76. Fridy, J.A., Orhan, C.: Statistical limit superior and limit inferior. Proc. Amer. Math. Soc. 125, 3625–3631 (1997)

77. Gadjiev, A.D.: The convergence problem for a sequence of positive linear operators on unbounded sets, and theorems analogous to that of P.P. Korovkin. Soviet Math. Dokl. 15, 1433–1436 (1974)

78. Gadjiev, A.D.: Linear k-positive operators in a space of regular functions and theorems of P.P. Korovkin type (Russian). Izv. Nauk Azerbajian SSR Ser. Fiz.-Tehn. Mat. Nauk 5, 49–53 (1974)

79. Gadjiev, A.D.: On P.P. Korovkin type theorems (Russian). Math. Zametki 20, 781–786 (1976)

80. Gadjiev, A.D., Orhan, C.: Some approximation theorems via statistical convergence. Rocky Mountain J. Math. 32, 129–138 (2002)

81. Gal, S.G.: Degree of approximation of continuous functions by some singular integrals. Rev. Anal. Numér. Théor. Approx. (Cluj) 27, 251–261 (1998)

82. Gal, S.G.: Approximation theory in fuzzy setting. In: Handbook of Analytic-Computational Methods in Applied Mathematics, pp. 617–666. Chapman & Hall/CRC, Boca Raton, FL (2000)

83. Gal, S.G.: Shape-Preserving Approximation by Real and Complex Polynomials. Birkhäuser Boston, Inc., Boston (2008)

84. Goetschel, R.J., Voxman, W.: Elementary fuzzy calculus. Fuzzy Sets and Systems 18, 31–43 (1986)

85. Grauert, H., Fritzsche, K.: Several Complex Variables (Translated from the German). Graduate Texts in Mathematics, vol. 38. Springer, Heidelberg (1976)

86. Hardy, G.H.: Divergent Series. Oxford Univ. Press, London (1949)

87. İspir, N.: Convergence of sequences of k-positive linear operators in subspaces of the space of analytic functions. Hacet. Bull. Nat. Sci. Eng. Ser. B 28, 47–53 (1999)

88. Kaleva, O.: Fuzzy differential equations. Fuzzy Sets and Systems 24, 301–317 (1987)

89. Karakuş, S., Demirci, K., Duman, O.: Equi-statistical convergence of positive linear operators. J. Math. Anal. Appl. 339, 1065–1072 (2008)

90. Karakuş, S., Demirci, K., Duman, O.: Statistical convergence on intuitionistic fuzzy normed spaces. Chaos, Solitons & Fractals 35, 763–769 (2008)

91. Knoop, H.B., Pottinger, P.: Ein satz von Korovkin-typ für C^k-raüme. Math. Z. 148, 23–32 (1976)

92. Kolk, E.: Matrix summability of statistically convergent sequences. Analysis 13, 77–83 (1993)

93. Korovkin, P.P.: Linear Operators and Approximation Theory. Hindustan Publ. Corp., Delhi (1960)

94. Krantz, S.G.: Function Theory of Several Complex Variables (Reprint of the 1992 edition). AMS Chelsea Publishing, Providence (2001)

95. Liu, P.: Analysis of approximation of continuous fuzzy functions by multivariate fuzzy polynomials. Fuzzy Sets and Systems 127, 299–313 (2002)

96. Maddox, I.J.: Statistical convergence in a locally convex space. Math. Proc. Cambridge Phil. Soc. 104, 141–145 (1988)

97. Matloka, M.: Sequences of fuzzy numbers. BUSEFAL 28, 28–37 (1986)

98. Miller, H.I.: Rates of convergence and topics in summability theory. Akad. Nauka Umjet. Bosne Hercegov. Rad. Odjelj. Prirod. Mat. Nauka 22, 39–55 (1983)

99. Miller, H.I.: A measure theoretical subsequence characterization of statistical convergence. Trans. Amer. Math. Soc. 347, 1811–1819 (1995)

100. Miller, K.S., Ross, B.: An Introduction to The Fractional Calculus and Fractional Differential Equations. John Wiley & Sons, Inc., New York (1993)

101. Niven, I., Zuckerman, H.S., Montgomery, H.: An Introduction to the Theory of Numbers, 5th edn. Wiley, New York (1991)

102. Nuray, F., Savaş, E.: Statistical convergence of sequences of fuzzy numbers. Math. Slovaca 45, 269–273 (1995)

103. Oldham, K.B., Spanier, J.: The Fractional Calculus: Theory and Applications of Differentiation and Integration to Arbitrary Order. Mathematics in Science and Engineering, vol. 111. Academic Press, New York (1974)

104. Özarslan, M.A., Duman, O., Doğru, O.: Rates of A-statistical convergence of approximating operators. Calcolo 42, 93–104 (2005)

105. Özarslan, M.A., Duman, O., Srivastava, H.M.: Statistical approximation results for Kantorovich-type operators involving some special functions. Math. Comput. Modelling 48, 388–401 (2008)

106. Šalat, T.: On statistically convergent sequences of real numbers. Math. Slovaca 30, 139–150 (1980)

107. Wu, C.X., Ma, M.: Embedding problem of fuzzy number space I. Fuzzy Sets and Systems 44, 33–38 (1991)

108. Zygmund, A.: Trigonometric Series. Cambridge Univ. Press, Cambridge (1979)

List of Symbols

Index